Tharwat F. Tadros
Emulsions
De Gruyter Graduate

Also of Interest

Nanodispersions
Tadros, 2015
ISBN 978-3-11-029033-2, e-ISBN 978-3-11-029034-9

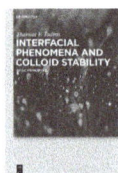

Interfacial Phenomena and Colloid Stability:
Volume 1 Basic Principles
Tadros, 2015
ISBN 978-3-11-028340-2, e-ISBN 978-3-11-028343-3

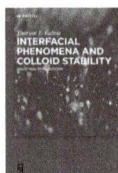

Interfacial Phenomena and Colloid Stability:
Volume 2 Industrial Applications
Tadros, 2015
ISBN 978-3-11-037107-9, e-ISBN 978-3-11-036647-1

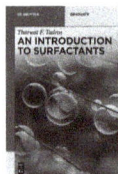

An Introduction to Surfactants
Tadros, 2014
ISBN 978-3-11-031212-6, e-ISBN 978-3-11-031213-3

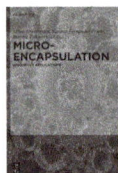

Microencapsulation: Innovative Applications
Giamberini, Fernandez-Prieto, Tylkowski (Eds.), 2015
ISBN 978-3-11-033187-5, e-ISBN 978-3-11-033199-8

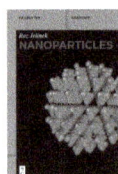

Nanoparticles
Jelinek, 2015
ISBN 978-3-11-033002-1, e-ISBN 978-3-11-033003-8

Tharwat F. Tadros

Emulsions

Formation, Stability, Industrial Applications

DE GRUYTER

Author
Prof. Tharwat F. Tadros
89 Nash Grove Lane
Workingham RG40 4HE
Berkshire, UK
tharwat@tadros.fsnet.co.uk

ISBN 978-3-11-045217-4
e-ISBN (PDF) 978-3-11-045224-2
e-ISBN (EPUB) 978-3-11-045226-6

Library of Congress Cataloging-in-Publication Data
A CIP catalog record for this book has been applied for at the Library of Congress.

Bibliographic information published by the Deutsche Nationalbibliothek
The Deutsche Nationalbibliothek lists this publication in the Deutsche Nationalbibliografie;
detailed bibliographic data are available on the Internet at http://dnb.dnb.de.

© 2016 Walter de Gruyter GmbH, Berlin/Boston
Cover image: surachetkhamsuk/iStock/thinkstock
Typesetting: PTP-Berlin, Protago-TEX-Production GmbH, Berlin
Printing and binding: CPI books GmbH, Leck
♾ Printed on acid-free paper
Printed in Germany

www.degruyter.com

Preface

Emulsions of disperse systems consist of oil, water and an emulsifier (usually a surfactant). If the disperse phase is oil and the continuous phase is water, the system is referred to as an oil/water (O/W) emulsion, whereas if the disperse phase is water and the continuous phase is oil the system is referred to as a W/O emulsion. A special class is where the disperse phase is, say, a polar oil and the continuous phase is oil and vice versa is referred to as an oil/oil (O/O) emulsion. All emulsions are thermodynamically unstable, since the free energy of formation is positive (the increase in interfacial energy due to expansion of the interface which is positive is higher than the positive entropy of dispersion). Thus the emulsion only becomes kinetically stable in the presence of an emulsifier. The latter produces an energy barrier as a result of the repulsive interaction that overcomes the van der Waals attraction. This energy barrier prevents the breakdown of the emulsion by flocculation and coalescence. The higher the energy barrier, the longer the shelf life of the emulsion. In order to understand the process of emulsification one must consider the process of the deformation of droplets and their break-up into smaller droplets. An important aspects that must be considered is the Laplace pressure of the drop (given by $2\gamma/R$, where γ is the interfacial tension and R is the drop radius). On deformation of the drop, the Laplace pressure increases, and energy must be applied to overcome this increase of the pressure. Surfactants play a major role in emulsion formation by lowering the interfacial tension. In addition, the surfactant prevents coalescence of the droplets by producing an interfacial tension and Gibbs elasticity. The surfactant molecules diffuse to the uncovered area of the interface, carrying liquid with them, thus preventing thinning and disruption of the liquid between the droplets (the Marangoni effect). This prevents coalescence of the droplets during emulsification. It is important to consider the process of surfactant adsorption under dynamic and equilibrium conditions. Several methods can be applied for emulsification, mostly the use of high speed stirrers (rotor-stator mixers), homogenisers and ultrasonic techniques. Several semi-empirical methods can be applied for selection of the emulsifier of which the hydrophilic-lipophilic-balance (HLB) and the phase inversion temperature principles are the most commonly used in practice. A more quantitative method for emulsifier selection is the cohesive energy ratio concept (CER), which takes into account the various interactions that occur between the lipophilic part of the surfactant at the oil side and the hydrophilic part of the molecule at the water side. Another useful concept for emulsifier selection is using the critical packing parameter concept, which describes the shape of the surfactant molecule and the ratio between the cross sectional area of the lipophilic part of the molecule and the hydrophilic head group. The various processes of emulsion breakdown, namely creaming or sedimentation, flocculation, Ostwald ripening, coalescence and phase inversion, must be considered at a fundamental level, highlighting the various methods that can be applied to prevent each one of these breakdown processes. Creaming or sedimentation is the result of gravity when the gravitational force (which is given by the product of the mass and acceleration due to gravity) exceeds the Brownian diffusion (given by kT, where k is the Boltzmann constant and T is the absolute temperature). Theories are available for calculating the rate of creaming or sedimentation in very dilute emulsions (with a volume fraction $\varphi \leq 0.01$) and moderately concentrated emulsions ($0.2 > \varphi > 0.01$). With more concentrated emulsions ($\varphi > 0.2$), the rate of creaming or sedimentation is a complex function of the volume fraction, and only semi-empirical equations are available to describe the rate in such concentrated emulsions. For prevention of creaming or sedimentation, two main methods are possible: (i) reduction of droplet size (R < 100 nm) such that the Brownian diffusion overcomes the gravity force; (ii) addition of thickeners (mostly high molecular weight polymers) that produce a high viscosity at very low shear rates (residual or zero shear viscosity $\eta(0)$) and low viscosity at high shear rates (shear thinning systems). The high $\eta(0)$ reduces the rate of creaming or sedimentation which can become close to zero. Flocculation of emulsions can be eliminated by introducing a strong repulsive energy. The latter can be produced by electrostatic repulsion, e.g. in the presence of an ionic surfactant and/or steric repulsion in the presence of adsorbed non-ionic or polymeric surfactants. Ostwald ripening results from the difference in solubility between small and large droplets. The smaller droplets have higher solubility than the larger ones. On storage, molecular diffusion of the oil molecules occurs from the smaller to the larger droplets. With time the droplet size distribution shifts to larger values. These larger droplets enhance creaming or sedimentation, flocculation and coalescence. Ostwald ripening can be reduced by the addition of a small percentage of highly insoluble oil (such as squalane) and modification of the interfacial film by enhancing the Gibbs elasticity. Coalescence results from thinning and disruption of the liquid film between the droplets on their close approach, in a cream layer, in a floc or during Brownian collision. A useful concept describing the process of coalescence was introduced by Deryaguin, who introduced the concept of disjoining pressure $\pi(h)$. The latter is simply the difference in pressure between a thin film with thickness h, $P(h)$ and that of an infinitely thick film P_0. $\pi(h)$ is made from three contributions: an attractive term, π_{VDW} (which is negative) and two repulsive terms, namely electrostatic π_E, and steric π_s, both of which are positive. To obtain a stable film $\pi_E + \pi_s \gg \pi_{VDW}$, and this can be achieved by using mixed surfactant films, lamellar liquid crystalline phases or polymeric surfactants. Phase inversion can be of two types: (i) catastrophic, for example by increasing the volume fraction of the disperse phase above a critical value (usually above 0.6–0.7); (ii) transitional inversion, produced by changing the emulsion condition such as increase of temperature and/or addition of electrolyte. Unfortunately there is no theory that can explain phase inversion with a fundamental approach, and methods must be adopted to prevent it from happening.

This book is organised as follows:

Chapter 1 gives a general introduction, starting with the definition of emulsions and the role of the emulsifier. Classification of emulsions can be based on the nature of the emulsifier or the structure of the system. A brief description of the general instability problems with emulsions is given: creaming/sedimentation, flocculation, Ostwald ripening, coalescence and phase inversion.

Chapter 2 describes the thermodynamics of emulsion formation and breakdown. The second law of thermodynamics is applied for emulsion formation, namely the balance of energy and entropy and non-spontaneous formation of emulsions. The breakdown of the emulsion by flocculation and coalescence in the absence of an emulsifier is described. This is followed by a description of the role of the emulsifier in preventing flocculation and coalescence by creating an energy barrier resulting from the repulsive energies between the droplets.

Chapter 3 describes the interaction forces between emulsion droplets. First, the van der Waals attraction and its dependence on droplet size, Hamaker constant and separation distance between the droplets is given. This is followed by an analysis of the electrostatic repulsion resulting from the presence of electrical double layers and its dependence on surface (or zeta) potential and electrolyte concentration and valency. Combination of the van der Waals attraction with double-layer repulsion forms the basis of the theory of colloid stability. The steric repulsion resulting from the presence of adsorbed non-ionic surfactants and polymers is described. Combination of van der Waals attraction with steric repulsion and the theory of steric stabilisation is described at a fundamental lever.

Chapter 4 describes the adsorption of surfactants at the oil/water interface. The thermodynamic analysis of surfactant adsorption and the Gibbs adsorption isotherm are described. This is followed by calculation of the amount of surfactant adsorption and area per surfactant molecule at the interface. The various experimental techniques for measuring the interfacial tension are described.

Chapter 5 describes the mechanism of emulsification and the role of the emulsifier. The factors responsible for droplet deformation and its break-up are described at a fundamental level. The role of surfactants in preventing coalescence during emulsification are described in terms of the Gibbs dilational elasticity and the Marangoni effect.

Chapter 6 gives a brief description of methods of emulsification, namely pipe flow, static mixers and high speed stirrers (rotor-stator mixer). The definition of laminar and turbulent flow is described in terms of the Reynolds number. Other emulsification methods such as membrane emulsification, high pressure homogenisers and ultrasonic methods are briefly described.

Chapter 7 describes the various methods that can be applied for selection of emulsifiers. The hydrophilic-lipophilic balance (HLB) and its application in surfactant selection is described, followed by methods of calculation of HLB numbers and the effect of the nature of the oil phase. The phase inversion temperature (PIT) method for emulsifier selection is described, followed by the cohesive energy ratio method for emulsifier selection. The final method for emulsifier selection, namely the critical packing parameter concept, is described.

Chapter 8 describes the process of creaming/sedimentation of emulsions and its prevention. The driving force for creaming/sedimentation, namely the effect of gravity, droplet size and density difference between the oil and continuous phase is described. This is followed by calculation of the rate of creaming/sedimentation in dilute emulsions and the influence of increase of the volume fraction of the disperse phase on the rate of creaming/sedimentation. The reduction of creaming/sedimentation is described in terms of the balance of the density of the two phases, reduction of droplet size and effect of addition of 'thickeners'.

Chapter 9 deals with the process of flocculation of emulsions and its prevention. The factors affecting flocculation are described, followed by calculation of fast and slow flocculation rate. This leads to a definition of stability ratio and its dependence on electrolyte concentration and valency. The definition of the critical coagulation concentration and its dependence on electrolyte valency is described.

Chapter 10 describes the process of Ostwald ripening and its reduction. The factors responsible for Ostwald ripening are described in terms of the difference in solubility between small and large droplets and application of the Kelvin equation. This is followed by calculation of the rate of Ostwald ripening and its experimental determination. The reduction of Ostwald ripening by incorporation of a small amount of highly insoluble oil and by the use of strongly adsorbed polymeric surfactant and enhancement of the Gibbs elasticity is described.

Chapter 11 deals with the process of emulsion coalescence and its prevention. The driving force for emulsion coalescence is described in terms of thinning and disruption of the liquid film between the droplets. The concept of disjoining pressure for prevention of coalescence is analysed. This is followed by the methods that can be applied for reduction or elimination of coalescence, namely the use of mixed surfactant films, the use of lamellar liquid crystalline phases and the use of polymeric surfactants.

Chapter 12 describes the process of phase inversion and its prevention. A distinction between catastrophic and transient phase inversion is given with particular reference to the influence of the disperse volume fraction and surfactant HLB number. An explanation is given for the factors responsible for phase inversion.

Chapter 13 summarises the various methods that can be applied for characterisation of emulsions. Particular attention is given to the methods of measurement of droplet size distribution, namely optical microscopy and image analysis, phase contrast and polarising microscopy, diffraction methods, confocal laser microscopy and back scattering methods.

Chapter 14 describes some industrial applications of emulsions, with particular emphasis on food, pharmacy, cosmetics, agrochemicals, rolling oils and lubricants.

Based on the above descriptions, it is clear that this book covers a wide-range of topics on emulsion formation, stability and applications. The various fundamental aspects involved in the various breakdown processes of emulsions are comprehensively analysed. It also highlights the aspects of emulsion production and their characterisation. In addition the book describes the various methods than can be applied to prevent all the breakdown processes of emulsions. It is hoped that this book will be of great help to emulsion research scientists in both academia and industry. It will also be extremely valuable for postgraduate students who are involved in this research area. It will also be very helpful to researchers just entering this field of research.

Tharwat Tadros   March, 2016

Contents

1 Emulsions: Formation, stability, industrial applications

1.1 General introduction

Emulsions are a class of disperse systems consisting of two immiscible liquids [1–4]. The liquid droplets (the disperse phase) are dispersed in a liquid medium (the continuous phase). Several classes may be distinguished: oil-in-water (O/W), water-in-oil (W/O) and oil-in-oil (O/O). The latter class may be exemplified by an emulsion consisting of a polar oil (e.g. propylene glycol) dispersed in a non-polar oil (paraffinic oil), and vice versa. To disperse two immiscible liquids one needs a third component, namely the emulsifier. The choice of the emulsifier is crucial in formation of the emulsion and its long term stability [1–4].

There are many examples one could quote of naturally occurring emulsions: milk and the O/W and W/O emulsions associated with oil bearing rocks are just two examples. Emulsion types can be classified on the basis of the nature of the emulsifier or the structure of the system as shown in Tab. 1.1.

Tab. 1.1: Classification of emulsions.

Nature of emulsifier	Structure of the system
– Simple molecules and ions	– Nature of internal and external phase: O/W, W/O
– Nonionic Surfactants	– Nanoemulsions
– Ionic surfactants	– Micellar emulsions (microemulsions)
– Surfactant mixtures	– Macroemulsions
– Nonionic Polymers	– Bilayer droplets
– Polyelectrolytes	– Double and Multiple Emulsions
– Mixed polymers and surfactants	– Mixed emulsions
– Liquid crystalline phases	
– Solid particles	

1.2 Nature of the Emulsifier

The simplest type is ions such as OH^-, which can be specifically adsorbed on the emulsion droplet, thus producing a charge. An electrical double layer can be produced which provides electrostatic repulsion. This has been demonstrated with very dilute O/W emulsions by removing any acidity. Clearly that process is not practical. The most effective emulsifiers are non-ionic surfactants, such as alcohol ethoxylates with the general formula $C_xH_{2x+1}-O-(CH_2-CH_2-O)_nH$, which can be used to emulsify oil in water or water in oil. In addition they can stabilise the emulsion against flocculation and coalescence. Ionic surfactants such as sodium dodecyl sulphate can also be used as emulsifiers (for O/W), but the system is sensitive to the presence of electrolytes. Surfactant mixtures, e.g. ionic and non-ionic, or mixtures of non-ionic surfactants, can be more effective in emulsification and stabilization of the emulsion. Non-ionic polymers, sometimes referred to as polymeric surfactants, e.g. pluronics, with the general formula $HO-(CH_2-CH_2-O)_n-(CH_2-CH(CH_3)-O)_m-(CH_2-CH_2-O)_n-OH$ or PEO–PPO–PEO, are more effective in stabilisation of the emulsion, but they may suffer from the difficulty of emulsification (to produce small droplets) unless high energy is applied for the process. Polyelectrolytes such as poly(methacrylic acid) can also be applied as emulsifiers. Mixtures of polymers and surfactants are ideal in achieving ease of emulsification and stabilisation of the emulsion. Lamellar liquid crystalline phases that can be produced using surfactant mixtures are very effective in emulsion stabilisation. Solid particles that can accumulate at the O/W interface can also be used for emulsion stabilisation. These are referred to as Pickering emulsions, whereby particles are partially wetted by the oil phase and partially by the aqueous phase.

1.3 Structure of the system

(i) Macroemulsions O/W and W/O: These usually have a size range of 0.1–5 μm with an average of 1–2 μm. These systems are usually opaque or milky due to the large size of the droplets and the significant difference in refractive index between the oil and water phases.

(ii) Nano-emulsions: These usually have a size range 20–100 nm. Like macroemulsions they are only kinetically stable. They can be transparent, translucent or opaque, depending on the droplet size, the refractive index difference between the two phases and the volume fraction of the disperse phase.

(iii) Double and multiple emulsions: these are emulsions-of-emulsions, W/O/W and O/W/O systems. They are usually prepared using a two-stage process. For example a W/O/W multiple emulsion is prepared by forming a W/O emulsion, which is then emulsified in water to form the final multiple emulsion.

(iv) Mixed emulsions: these are systems consisting of two different disperse droplet that do not mix in a continuous medium.

(v) Micellar emulsions or microemulsions: these usually have the size ranging from 5 to 50 nm. They are thermodynamically stable and strictly speaking they should not be described as emulsions. A better description is "swollen micelles" or "micellar systems".

The present book will only deal with macroemulsions, their formation, stability and industrial applications.

Several breakdown processes may occur on storage depending on: particle size distribution and density difference between the droplets and the medium; the magnitude of the attractive vs repulsive forces which determines flocculation; solubility of the disperse droplets and the particle size distribution which determines Ostwald ripening; stability of the liquid film between the droplets that determines coalescence; phase inversion, where the two phases exchange, e.g. an O/W emulsion inverting to W/O and vice versa. Phase inversion can be catastrophic, as in the case when the oil phase in an O/W emulsion exceeds a critical value. The inversion can be transient when for example the emulsion is subjected to temperature increase.

1.4 Breakdown processes in emulsions

The various breakdown processes are illustrated in the Fig. 1.1. The physical phenomena involved in each breakdown process is not simple, and it requires analysis of the various surface forces involved. In addition, the above processes may take place simultaneously rather then consecutively, and this complicates the analysis. Model emulsions with monodisperse droplets cannot be easily produced, and hence any theoretical treatment must take into account the effect of droplet size distribution. Theories that take into account the polydispersity of the system are complex, and in many cases only numerical solutions are possible. In addition, measurement of surfactant and polymer adsorption in an emulsion is not easy, and one has to extract such information from measurement at a planer interface.

Below a summary of each of the above breakdown processes is given, and details of each process and methods of its prevention is given in separate sections.

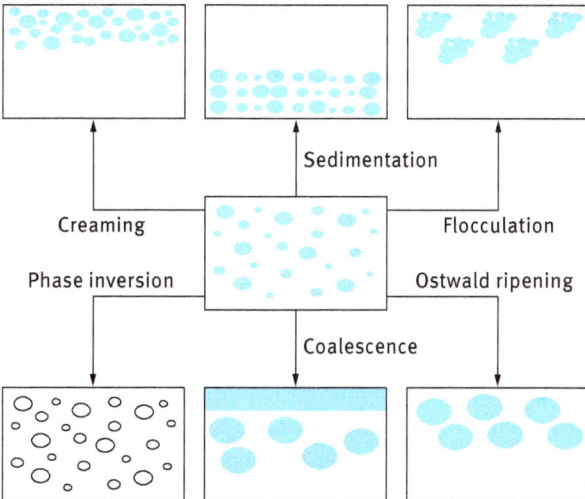

Fig. 1.1: Schematic representation of the various breakdown processes in emulsions.

1.5 Creaming and sedimentation

This process, with no change in droplet size, results from external forces usually gravitational or centrifugal. When such forces exceed the thermal motion of the droplets (Brownian motion), a concentration gradient builds up in the system, with the larger droplets moving faster to the top (if their density is lower than that of the medium) or to the bottom (if their density is larger than that of the medium) of the container. In the limiting cases, the droplets may form a close-packed (random or ordered) array at the top or bottom of the system, with the remainder of the volume occupied by the continuous liquid phase.

1.6 Flocculation

This process refers to aggregation of the droplets (without any change in primary droplet size) into larger units. It is the result of the van der Waals attraction, which is universal to all disperse systems. The main force of attraction arises from the London dispersion force that results from charge fluctuations of the atoms or molecules in the disperse droplets. The van der Waals attraction increases with a decrease in the distance separating the droplets, and at small separation distances the attraction becomes very strong, resulting in droplet aggregation or flocculation. The latter occurs when there is not enough repulsion to keep the droplets apart to distances where the van der Waals attraction is weak. Flocculation may be "strong" or "weak", depending on the magnitude of the attractive energy involved. In cases where the net attractive forces are relatively weak, an equilibrium degree of flocculation may be achieved (so-called weak flocculation), associated with the reversible nature of the aggregation process. The exact nature of the equilibrium state depends on the characteristics of the system. One can envision the build-up of aggregate-size distribution, and an equilibrium may be established between single droplets and large aggregates. With a strongly flocculated system, one refers to a system in which all the droplets are present in aggregates due to the strong van der Waals attraction between the droplets.

1.7 Ostwald ripening (disproportionation)

This results from the finite solubility of the liquid phases. Liquids which are referred to as being immiscible often have mutual solubilities which are not negligible. With emulsions which are usually polydisperse, the smaller droplets will have larger solubility compared to the larger ones (due to the effects of curvature). With time, the smaller droplets disappear, and their molecules diffuse to the bulk and become deposited on the larger droplets. With time the droplet size distribution shifts to larger values.

1.8 Coalescence

This refers to the process of thinning and disruption of the liquid film between the droplets which may be present in a creamed or sedimented layer, in a floc or simply during droplet collision, with the result of fusion of two or more droplets into larger ones. This process of coalescence results in a considerable change of the droplet size distribution, which shifts to larger sizes. The limiting case for coalescence is the complete separation of the emulsion into two distinct liquid phases. The thinning and disruption of the liquid film between the droplets is determined by the relative magnitudes of the attractive versus repulsive forces. To prevent coalescence, the repulsive forces must exceed the van der Waals attraction, thus preventing film rupture.

1.9 Phase inversion

This refers to the process whereby there will be an exchange between the disperse phase and the medium. For example, an O/W emulsion may with time or change of conditions invert to a W/O emulsion. In many cases, phase inversion passes through a transition state during which multiple emulsions are produced. For example, with an O/W emulsion, the aqueous continuous phase may become emulsified in the oil droplets, forming a W/O/W multiple emulsion. This process may continue until the entire continuous phase is emulsified into the oil phase, thus producing a W/O emulsion.

1.10 Industrial applications of emulsions

There are a number of industrial applications of emulsions worth noting: food emulsion, e.g. mayonnaise, salad creams, deserts, beverages etc.; personal care and cosmetics, e.g. hand creams, lotions, hair sprays, sunscreens, etc.; agrochemicals, e.g. self emulsifiable oils which produce emulsions on dilution with water, emulsion concentrates (EWs) and crop oil sprays; pharmaceuticals, e.g. anaesthetics of O/W emulsions, lipid emulsions, double and multiple emulsions, etc.; paints, e.g. emulsions of alkyd resins, latex emulsions, etc.; dry cleaning formulations, which may contain water droplets emulsified in the dry cleaning oil which is necessary to remove soils and clays; Bitumen emulsions – these are stable emulsions prepared in the containers, but when applied the road chippings they must coalesce to form a uniform film of bitumen; emulsions in the oil industry – many crude oils contain water droplets (for example North Sea oil), and these must be removed by coalescence followed by separation; oil slick dispersions – the oil spilled from tankers must be emulsified and then separated; emulsification of unwanted oil – this is an important process for pollution control.

The importance of emulsion in industry justifies a great deal of basic research to understand the origin of instability and methods to prevent their breakdown. Unfortunately fundamental research on emulsions is not easy, since model systems (e.g. with monodisperse droplets) are difficult to produce. In many cases, theories on emulsion stability are not exact, and semi-empirical approaches are used.

1.11 Book outline

Chapter 2 describes the thermodynamics of emulsion formation and breakdown. It begins with a section on definition of the interfacial region using the Gibbs concept of mathematical division of the interface. The application of the second law of thermodynamics leads to the definition of the interfacial tension which is given by the change of the Gibbs energy with area at constant temperature and composition (mJ m^{-1} or mN m^{-1}) The application of the second law of thermodynamics for emulsion formation and breakdown is given by the balance of interfacial energy and entropy, and this shows that the free energy of the formation of emulsions is positive, which explains the non-spontaneous formation of emulsions. The breakdown of the emulsion by flocculation, Ostwald ripening and coalescence in the absence of an emulsifier is the result of a reduction in interfacial energy due to expansion of the interface. The role of the emulsifier in preventing flocculation and coalescence by creating an energy barrier resulting from the repulsive energies between the droplets is described. Chapter 3 describes the interaction forces between emulsion droplets. It starts with Van der Waals attraction and its dependence on droplet size, Hamaker constant and separation distance between the droplets. This attraction is prevented by electrostatic repulsion resulting from the presence of electrical double layers and its dependence on surface (or zeta) potential and electrolyte concentration and valency. Combination of the van der Waals attraction with double layer repulsion and the theory of colloid stability is described. The steric repulsion resulting from the presence of adsorbed non-ionic surfactants and polymers is described in terms of the unfavourable mixing of the adsorbed layers (when these are in good solvent conditions) and loss of configurational entropy on significant overlap of the adsorbed layers. Combination of van der Waals attraction with steric repulsion describes the theory of steric stabilisation. The main factors responsible for effective steric stabilisation are described. Chapter 4 describes the adsorption and orientation of surfactants molecules at the oil/water (O/W) interface. Thermodynamic analysis of surfactant adsorption is analysed using the second law of thermodynamics and this results in the Gibbs adsorption isotherm. The calculation of the amount of surfactant adsorption and area per surfactant molecule at the interface from the variation of interfacial tension with surfactant concentration are described. The results show the dependence of surfactant adsorption on the nature of surfactant molecule, namely the length of the hydrocarbon chain and the nature of the head group. The experimental techniques for measuring the interfacial tension are briefly described. Chapter 5 deals with the mechanism of emulsification and the role of the emulsifier. It starts with description of the factors responsible for droplet deformation and its break-up. The role of surfactant in preventing coalescence during emulsification is described in terms of the Gibbs dilational elasticity and the Marangoni effect. Chapter 6 describes the methods that can be applied for emulsification. Low energy emulsification methods are described in terms of pipe flow and static mixers. Medium energy methods use high speed stirrers such as the rotor-stator mixer. High energy emulsification involves the use of High pressure homogenisers and ultrasonic techniques. A distinction between laminar and turbulent flow can be described in terms of the dimensionless Reynold's number. Chapter 7 deals with the methods that can be applied for selection of emulsifiers. A semi-empirical method is based on the hydrophilic-lipophilic-balance (HLB) which can be easily calculated for simple surfactant molecules such as non-ionic surfactants based on alkyl polyethylene oxide. For more complex molecules a method based on group numbers can be applied. The dependence of the HLB number on the nature of the oil is described. The experimental method that can be applied for selecting the optimum HLB number for a given oil is described together with its application in surfactant selection. The phase inversion temperature (PIT) method for emulsifier selection based on polyethylene oxide is described. Particular attention is given on the variation of interfacial tension with temperature which shows a minimum at the PIT. By preparing the emulsion at temperatures close to the PIT, very small droplets can be produced which can be stabilised against coalescence by rapid cooling of the emulsion. The cohesive energy ratio (CER) method for emulsifier selection based on the dispersing tendencies at the oil and water interfaces of the emulsifier molecule is described. The procedure for calculating the CER is described. Chapter 8 deals with the process of creaming/sedimentation of emulsions and its prevention. The driving force for creaming/sedimentation, namely the effect of gravity, droplet size and density difference between the oil and continuous phase are discussed at a fundamental level. The calculation of the rate of creaming/sedimentation in dilute emulsions is described using Stokes' law. The influence of increase of the volume fraction of the disperse phase on the rate of creaming/sedimentation is described. The various methods that can be applied for reduction of creaming/sedimentation are described : Balance of the density of the two phases, reduction of droplet size and effect of addition of 'thickeners'. Chapter 9 describes the flocculation of emulsions and its prevention. The factors responsible for flocculation are described. This is followed by the theories of calculation of fast and slow flocculation rates, k_o and k respectively. The stability ratio W is described by the ratio of k_o/k, and the dependence of W on electrolyte concentration and valency is described. This leads to the definition of the critical coagulation concentration and its dependence on electrolyte valency. The reduction of flocculation by enhancing the repulsive forces is described. Chapter 10 deals with the process of Ostwald ripening and its reduction. The factors responsible for Ostwald ripening, namely the difference in solubility between small and large droplets as described by the Kelvin and Ostwald equations are described. The theory of calculation of the rate of Ostwald ripening is given. The methods that can be applied for reduction of Ostwald ripening are described: incorporation of a small amount of highly insoluble oil and the use of strongly adsorbed polymeric surfactant which enhances the Gibbs elasticity. Chapter 11 describes the process of emulsion coalescence and its prevention. The driving force for emulsion coalescence in terms of thinning and disruption of the liquid film between the droplets is described. The concept of disjoining pressure for describing the process of coalescence prevention is described. The methods for reduction or elimination of coalescence are briefly described: use of mixed surfactant films, use of lamellar liquid crystalline phases and use of polymeric surfactants. Chapter 12 describes the process of phase inversion and its prevention. A distinction can be made between catastrophic and transient phase inversion. Catastrophic phase inversion can be produced by increasing the disperse phase volume fraction above its maximum packing. Transient phase inversion can occur when the surfactant properties change with factors such as increase of temperature and/or addition of electrolyte. Chapter 13 describes the various techniques that can be applied for characterisation of emulsions, namely measurement of droplet size distribution by optical microscopy and image analysis, phase contrast and polarising microscopy, diffraction methods, confocal laser microscopy and back scattering methods. Chapter 14 gives some examples of industrial applications of emulsions, namely in the food industry, pharmacy, cosmetics and personal care, agrochemicals, and in rolling oil emulsions and lubricants.

References

[1] Th. F. Tadros and B. Vincent, in: P. Becher (ed.), *Encyclopedia of Emulsion Technology*, Marcel Dekker, New York, 1983.
[2] B. P. Binks (ed.), *Modern Aspects of Emulsion Science*, The Royal Society of Chemistry Publication, Cambridge, 1998.
[3] Tharwat Tadros. *Applied Surfactants*, Wiley-VCH, Germany, 2005.
[4] Tharwat Tadros, Emulsion Formation Stability and Rheology, in: Tharwat Tadros (ed.), *Emulsion Formation and Stability*, Ch. 1, Wiley-VCH, Germany, 2013.

2 Thermodynamics of emulsion formation and breakdown

2.1 The interface (Gibbs dividing line)

When two immiscible phases α and β (oil and water) come into contact, an interfacial region develops. The interfacial region is not a layer which is one molecule thick; it is a region with thickness δ with properties different from the two bulk phases α and β. In bringing phases α and β into contact, the interfacial regions of these phases undergo some changes, resulting in a concomitant change in the internal energy. If we were to move a probe from the interior of α to that of β, one would, at some distance from the interface, begin to observe deviations in composition, in density and in structure; the closer to phase β the larger the deviations, until eventually the probe arrives in the homogeneous phase β. The thickness of the transition layer will depend on the nature of the interfaces and on other factors. Gibbs [1] considered the two phases α and β to have uniform thermodynamic properties up to the interfacial region. He assumed a mathematical plane Z^σ in the interfacial region in order to define the interfacial tension γ. A schematic representation of the interfacial region and the Gibbs mathematical plane is given in Fig. 2.1.

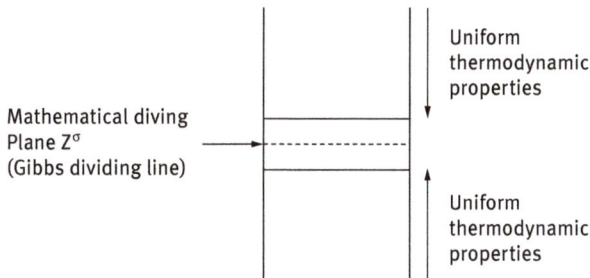

Uniform thermodynamic properties

Mathematical diving Plane Z^σ (Gibbs dividing line)

Uniform thermodynamic properties

Fig. 2.1: The Gibbs diving line.

Using the Gibbs model, it is possible to obtain a definition of the interfacial tension γ. The surface free energy dG^σ is made up of three components: an entropy term, $S^\sigma \, dT$; an interfacial energy term, $A \, d\gamma$; and a composition term, $\sum n_i \, d\mu_i$ (n_i is the number of moles of component i with chemical potential μ_i).

The Gibbs–Deuhem equation is

$$dG^\sigma = -S^\sigma \, dT + A \, d\gamma + \sum n_i \, d\mu_i . \tag{2.1}$$

At constant temperature and composition,

$$dG^\sigma = A\,d\gamma$$

$$\gamma = \left(\frac{\partial G^\sigma}{\partial A}\right)_{T,n_i}$$ (2.2)

For a stable interface, γ is positive, i.e. if the interfacial area increases G^σ increases. Note that γ is energy per unit area $(mJ\,m^{-2})$, which is dimensionally equivalent to force per unit length $(mN\,m^{-1})$, the unit usually used to define surface or interfacial tension.

An alternative approach to Gibbs treatment was given by Guggenheim [2] in which two dividing planes are drawn, one in phase α, the other in phase β as is illustrated in Fig. 2.2.

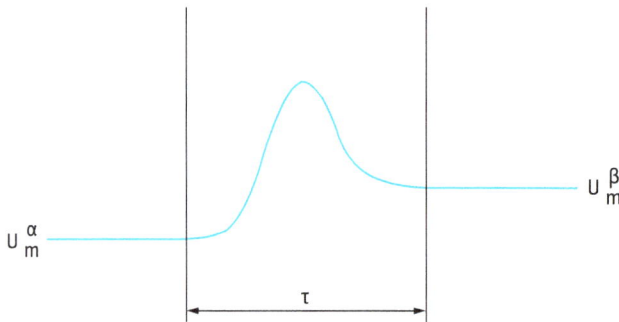

Fig. 2.2: Guggenheim convention of the interfacial region.

The planes are sufficiently far outside the interfacial region that beyond them the two phases have their bulk properties with internal energy U_m^α and U_m^β. In this convention the interfacial region is enclosed between two planes, a distance τ apart. The interfacial region has now a finite volume $A\tau$. In this convention the value of excess quantities depends on the choice of the two planes, and hence on τ, whereas in the Gibbs convention it depends on the choice of the location of the dividing plane. The analysis of Guggenheim [2] for the interfacial region is quite complex and hence the more simple Gibbs convention [1] is used in the analysis of interfacial phenomena.

For a curved interface, one should consider the effect of the radius of curvature. Fortunately, γ for a curved interface is estimated to be very close to that of a planer surface, unless the droplets are very small $(< 10\,nm)$.

Curved interfaces produce some other important physical phenomena which affect emulsion properties, e.g. the Laplace pressure Δp which is determined by the radii of curvature of the droplets,

$$\Delta p = \gamma\left(\frac{1}{r_1} + \frac{1}{r_2}\right),$$ (2.3)

where r_1 and r_2 are the two principal radii of curvature.

For a perfectly spherical droplet $r_1 = r_2 = r$, and

$$\Delta p = \frac{2\gamma}{r} \tag{2.4}$$

For a hydrocarbon droplet with radius 100 nm, and $\gamma_{O/W} = 50\,\text{mN m}^{-1}$, $\Delta p \sim 10^6$ Pa (~ 10 atm).

2.2 Thermodynamics of emulsion formation and breakdown

Consider a system in which an oil is represented by a large drop 2 of area A_1 immersed in a liquid 2, which is now subdivided into a large number of smaller droplets with total area A_2 ($A_2 \gg A_1$), as shown in Fig. 2.3. The interfacial tension γ_{12} is the same for the large and smaller droplets, since the latter are generally in the region of 0.1 to a few μm.

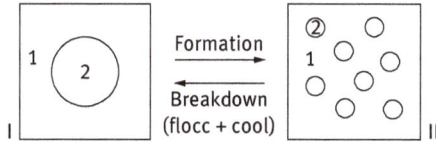

Fig. 2.3: Schematic representation of emulsion formation and breakdown.

The change in free energy in going from state I to state II is made from two contributions: a surface energy term (that is positive) that is equal to $\Delta A \gamma_{12}$ (where $\Delta A = A_2 - A_1$); and an entropy of dispersions term which is also positive (since producing a large number of droplets is accompanied by an increase in configurational entropy) which is equal to $T\Delta S^{\text{conf}}$.

From the second law of thermodynamics,

$$\Delta G^{\text{form}} = \Delta A \gamma_{12} - T\Delta S^{\text{conf}} . \tag{2.5}$$

In most cases $\Delta A \gamma_{12} \gg T\Delta S^{\text{conf}}$, which means that ΔG^{form} is positive, i.e. the formation of emulsions is non-spontaneous, and the system is thermodynamically unstable. In the absence of any stabilization mechanism, the emulsion will break by flocculation and coalescence, as illustrated in Fig. 2.4 by the full line. In this case there are no free-energy barriers either to flocculation or coalescence. The kinetics of both breakdown processes is diffusion controlled: in the case of flocculation, by the diffusion of the droplets, and in the case of coalescence, by diffusion of molecules of liquid 1 out of the thin liquid film formed between two contacting droplets of liquid 2. The dashed line in Fig. 2.4 corresponds to the case where sedimentation or creaming is superimposed upon the flocculation and coalescence. The final state of the system (state III) is now the more familiar one of two liquid phases separated by a flat interface. The dotted line in Fig. 2.4 represents the situation if, in addition to the above effects, Ostwald ripening

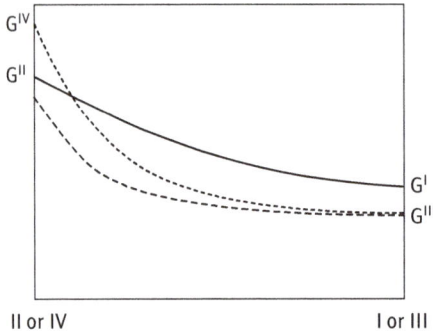

Fig. 2.4: Free energy path in emulsion breakdown: ——, Flocc. + coal., – – –, Flocc. + coal. + Sed., -------, Flocc. + coal. + sed. + Ostwald ripening.

has to be taken into account. This occurs if the initial state IV is polydisperse, and the liquids have a finite mutual solubility.

In the presence of a stabilizer (surfactant and/or polymer), an energy barrier is created between the droplets and therefore the reversal from state II to state I becomes non-continuous as a result of the presence of these energy barriers. This is illustrated in Fig. 2.5. In the presence of the above energy barriers, the system becomes kinetically stable. Strictly speaking, the ΔG_{flocc} and ΔG_{coal} are activation free energies. The intermediate state V is a metastable state and represents a flocculated emulsion that has not undergone any coalescence. If ΔG_{coal} is sufficiently high, it may stay in this state indefinitely.

Similarly, state II is also an unstable state, and if ΔG_{flocc} is sufficiently high, the stable, dispersed state may persist indefinitely. However, states II and V in these cases represent states of kinetic stability rather than true thermodynamic stability. The

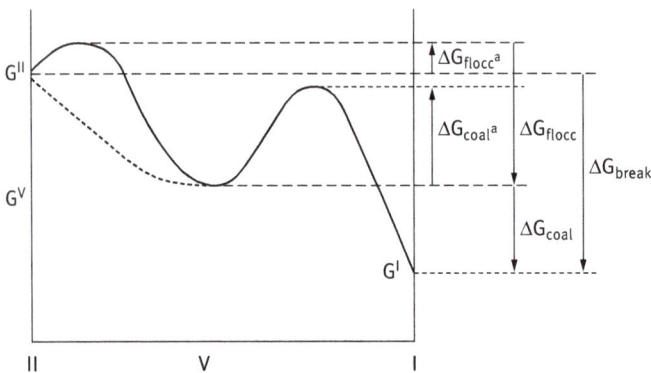

Fig. 2.5: Schematic representation of free energy path for breakdown (flocculation and coalescence) for systems containing an energy barrier.

dashed curve in Fig. 2.5 represents the situation if there is no free-energy barrier to flocculation, but where there is a large barrier to coalescence. Such a situation would arise for droplets stabilised, for example, by an adsorbed (neutral) polymer. In this case only the long-range van der Waals forces and the short-range steric repulsion forces (see Chapter 3) are operating. If ΔG_{flocc} is not too large (say < 10 kT per droplet), then the flocculation is reversible and an equilibrium is set-up, as will be discussed in Chapter 3.

From Fig. 2.5, it is seen that

$$\Delta G_{break} = \Delta G_{flocc} + \Delta G_{coal} . \tag{2.6}$$

It is worth considering the individual contributions to ΔG_{flocc} and ΔG_{coal} in the light of equations (2.5) and (2.6). The excess interfacial free energy G^{σ} associated with the presence of an interface is given by

$$G^{\sigma} = \Delta A \gamma_{12} + \sum_i \mu_i n_i^{\sigma} . \tag{2.7}$$

If an interface disappears due to coalescence, the change in free energy ΔG_{coal} is simply given by

$$\Delta G_{coal} = -\Delta(\gamma_{12}\Delta A) . \tag{2.8}$$

The term $\sum_i \mu_i n_i^{\sigma}$ disappears, since the chemical potential of species I is the same in either bulk phase and in the interface. Considering equations (2.5), (2.6) and (2.8) leads to the conclusion that

$$\Delta G_{flocc} = \Delta A \Delta \gamma_{12} - T \Delta S^{conf} . \tag{2.9}$$

Since

$$\Delta(\Delta A \gamma_{12}) = \gamma_{12}\Delta \Delta A + \Delta A \Delta \gamma_{12} , \tag{2.10}$$

then ΔG_{flocc} is made up of two terms: the $\Delta A \Delta \gamma_{12}$ associated with the change in interfacial tension in the contact region of two droplets (i.e., for the two surfaces in the film separating the droplets), and the $T \Delta S^{conf}$ term associated with the change in configurational entropy. Both terms are negative, and in most cases the $\Delta A \Delta \gamma_{12}$ term dominates, so that ΔG_{flocc} is negative, i.e., flocculation is thermodynamically spontaneous. However if ($\Delta A \Delta \gamma_{12}$) is less than ($T \Delta S^{conf}$), then ΔG_{flocc} is positive, and the emulsion is then thermodynamically stable against flocculation. This situation is schematically illustrated in Fig. 2.6.

This means that flocculation will not occur, and the emulsion has to be concentrated by creaming/sedimentation or centrifugation before coalescence can occur. The condition $|\Delta A \Delta \gamma_{12}| < |T \Delta S^{conf}|$ may be realised if $\Delta \gamma_{12}$ is small, i.e., the secondary minimum in the energy-distance curve (see Chapter 3) is small. Since $|T \Delta S^{conf}|$ decreases as the droplet number concentration increases, one can envision that at some initial droplet concentration $\Delta G_{flocc} = 0$, i.e. below this concentration the emulsion

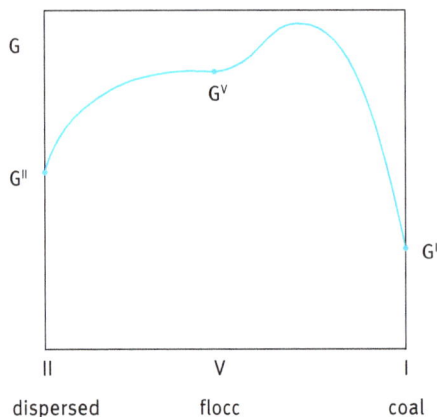

Fig. 2.6: Free-energy path for an emulsion which is thermodynamically stable with respect to flocculation.

is thermodynamically stable (ΔG_{flocc} positive), but that beyond this concentration the emulsion becomes thermodynamically unstable (ΔG_{flocc} negative) and reversible flocculation occurs.

References

[1] J. W. Gibbs, *Collected Papers, vol. 1, Thermodynamics*, Dover, USA, 1961.

[2] E. A. Guggenheim, *Thermodynamics*, 4th ed., North Holland, USA, 1959.

[3] Th. F. Tadros and B. Vincent, in: P. Becher (ed.), *Encyclopedia of Emulsion Technology*, Marcel Dekker, New York, 1983.

[4] Tharwat Tadros, *Applied Surfactants*, Wiley-VCH, Germany, 2005.

[5] B. P. Binks (ed.), *Modern Aspects of Emulsion Science*, The Royal Society of Chemistry Publication, Cambridge, 1998.

[6] Tharwat Tadros, Emulsion Formation Stability and Rheology, in: Tharwat Tadros (ed.), *Emulsion Formation and Stability*, Ch. 1, Wiley-VCH, Germany, 2013.

[7] Tharwat Tadros (ed.), *Encyclopedia of Colloid and Interface Science*, Springer, Germany, 2013.

3 Interaction forces between emulsion droplets

Generally speaking, there are three main interaction forces (energies) between emulsion droplets, which are discussed in the following.

3.1 Van der Waals attraction

As is well known, atoms or molecules always attract each other at short distances of separation. The attractive forces are of three different types: dipole–dipole interaction (Keesom), dipole-induced dipole interaction (Debye) and London dispersion force. The London dispersion force is the most important, since it occurs for polar and non-polar molecules. It arises from fluctuations in the electron density distribution.

At small distances of separation r in vacuum, the attractive energy between two atoms or molecules is given by

$$G_{aa} = -\frac{\beta_{11}}{r^6} , \tag{3.1}$$

where β_{11} is the London dispersion constant.

For emulsion droplets which are made of atom or molecular assemblies, the attractive energies have to be compounded. In this process, only the London interactions must be considered, since large assemblies have neither a net dipole moment nor a net polarization. The result relies on the assumption that the interaction energies between all molecules in one particle with all the other are simply additive [1]. The interaction between two identical spheres in vacuum the result is

$$G_A = -\frac{A_{11}}{6} \left(\frac{2}{s^2 - 4} + \frac{2}{s^2} + \ln \frac{s^2 - 4}{s^2} \right) . \tag{3.2}$$

A_{11} is known as the Hamaker constant and is defined by [1]:

$$A_{11} = \pi q_{11}^2 \beta_{ii} , \tag{3.3}$$

where q_{11} is number of atoms or molecules of type 1 per unit volume and s = (2R+h)/R. Equation (3.2) shows that A_{11} has the dimension of energy.

For very short distances (h ≪ R), equation (3.2) may be approximated by

$$G_A = -\frac{A_{11}R}{12h} . \tag{3.4}$$

When the droplets are dispersed in a liquid medium, the van der Waals attraction has to be modified to take into account the medium effect. When two droplets are brought from infinite distance to h in a medium, an equivalent amount of medium has to be transported the other way round. Hamaker forces in a medium are excess forces.

Consider two identical spheres 1 at a large distance apart in a medium 2, as illustrated in Fig. 3.1 (a). In this case the attractive energy is zero. Fig. 3.1 (b) gives the

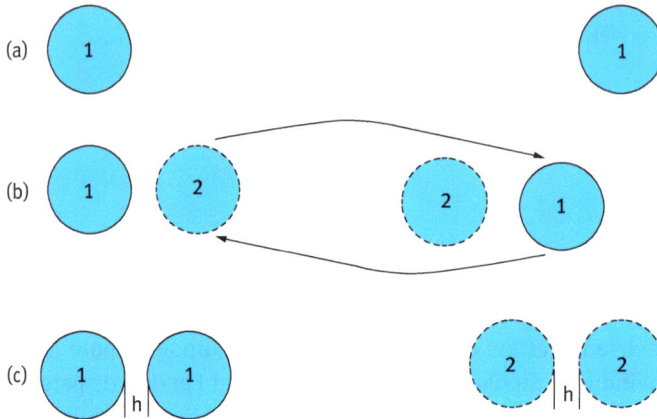

Fig. 3.1: Schematic representation of the interaction of two droplets in a medium.

same situation with arrows indicating the exchange of 1 against 2. Fig. 3.1 (c) shows the complete exchange which now shows the attraction between the two droplets 1 and 1 and equivalent volumes of the medium 2 and 2.

The effective Hamaker constant for two identical droplets 1 and 1 in a medium 2 is given by

$$A_{11(2)} = A_{11} + A_{22} - 2A_{12} = (A_{11}^{1/2} - A_{22}^{1/2})^2. \tag{3.5}$$

Equation (3.5) shows that two particles of the same material attract each other unless their Hamaker constant exactly matches each other. Equation (3.4) now becomes,

$$G_A = -\frac{A_{11(2)}R}{12h}, \tag{3.6}$$

where $A_{11(2)}$ is the effective Hamaker constant of two identical droplets with Hamaker constant A_{11} in a medium with Hamaker constant A_{22}.

In most cases the Hamaker constant of the droplets is higher than that of the medium. Examples of Hamaker constant for some liquids are given in Tab. 3.1. Generally speaking, the effect of the liquid medium is to reduce the Hamaker constant of the droplets below its value in vacuum (air).

G_A decreases with an increase of h, as schematically shown in Fig. 3.2. This shows the rapid increase of attractive energy with the decrease of h reaching a deep minimum at short h values. At extremely short h, the Born repulsion operates due to the overlap of the electronic clouds at such very small distance (a few Ångström). Thus in the absence of any repulsive mechanism, the emulsion droplets become strongly aggregated, due to the very strong attraction at short distances of separation.

To counteract the van der Waals attraction, it is necessary to create a repulsive force. Two main types of repulsion can be distinguished depending on the nature of the emulsifier used: Electrostatic (due to the creation of double layers) and Steric (due to the presence of adsorbed surfactant or polymer layers).

Tab. 3.1: Hamaker constant of some liquids.

Liquid	$A_{22} \cdot 10^{20}$ J
Water	3.7
Ethanol	4.2
Decane	4.8
Hexadecane	5.2
Cyclohexane	5.2

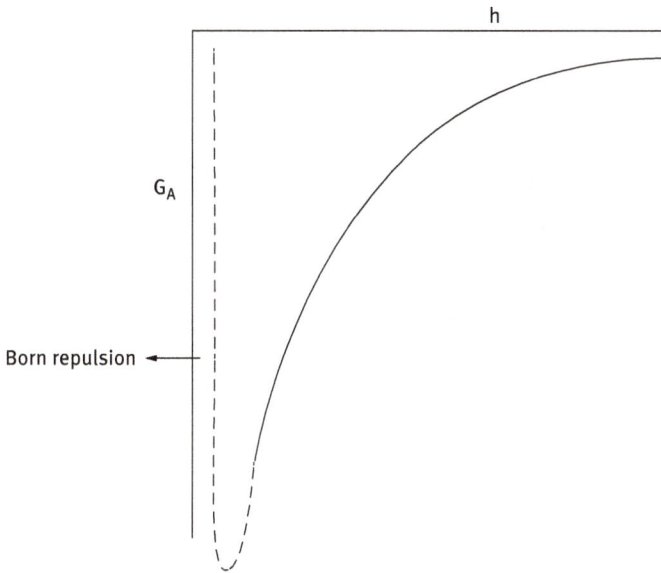

Fig. 3.2: Variation of G_A with h.

3.2 Electrostatic repulsion

This can be produced by adsorption of an ionic surfactant producing an electrical double layer whose structure was described by Gouy–Chapman, Stern and Grahame pictures [5]. A schematic representation of the diffuse double layer according to Gouy and Chapman [2] is shown in Fig. 3.3.

The surface charge σ_o is compensated by unequal distribution of counter ions (opposite in charge to the surface) and co-ions (same sign as the surface) which extend to some distance from the surface [2, 3]. The potential decays exponentially with distance x. At low potentials,

$$\psi = \psi_o \exp{-(\kappa x)} . \tag{3.7}$$

Note that when $x = 1/\kappa$, $\psi x = \psi_o/e - 1/\kappa$ is referred to as the "thickness of the double layer".

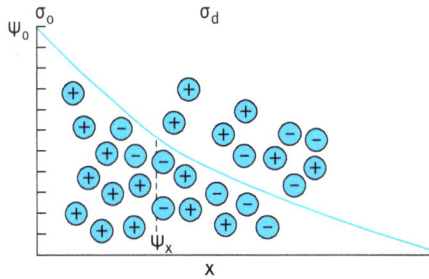

Fig. 3.3: Schematic representation of the diffuse double layer according to Gouy and Chapman [5].

The double layer extension depends on electrolyte concentration and valency of the counter ions:

$$\left(\frac{1}{\kappa} \right) = \left(\frac{\varepsilon_r \varepsilon_0 kT}{2n_0 Z_i^2 e^2} \right)^{1/2} . \tag{3.8}$$

ε_r is the permittivity (dielectric constant); 78.6 for water at 25 °C, ε_0 is the permittivity of free space, k is the Boltzmann constant and T is the absolute temperature, n_0 is the number of ions per unit volume of each type present in bulk solution and Z_i is the valency of the ions and e is the electronic charge.

Values of $(1/\kappa)$ at various 1:1 electrolyte concentrations are given in Tab. 3.2.

Tab. 3.2: Approximate values of $(1/\kappa)$ for 1:1 electrolyte (KCl).

C / mol dm^{-3}	10^{-5}	10^{-4}	10^{-3}	10^{-2}	10^{-1}
$(1/\kappa)$ / nm	100	33	10	3.3	1

The double layer extension increases with decrease in electrolyte concentration.

Stern [4] introduced the concept of the non-diffuse part of the double layer for specifically adsorbed ions, the rest being diffuse in nature. This is schematically illustrated in Fig. 3.4

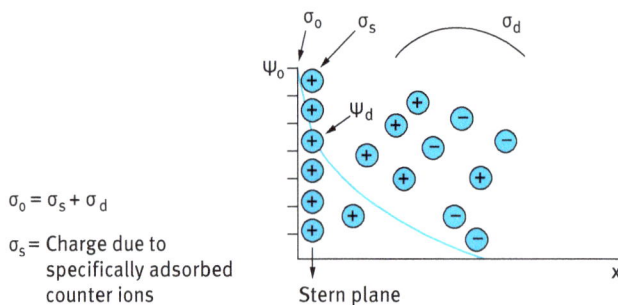

$\sigma_0 = \sigma_s + \sigma_d$

σ_s = Charge due to specifically adsorbed counter ions

Fig. 3.4: Schematic representation of the double layer according to Stern and Grahame [4, 5].

The potential drops linearly in the Stern region and then exponentially. Grahame [5] distinguished two types of ions in the Stern plane: physically adsorbed counterions (outer Helmholtz plane) and chemically adsorbed ions (that lose part of their hydration shell) (inner Helmholtz plane).

When charged droplets in an emulsion approach each other such that the double layers begin to overlap (droplet separation becomes less than twice the double layer extension), repulsion occurs. The individual double layers can no longer develop unrestrictedly, since the limited space does not allow complete potential decay [6]. This is illustrated in Fig. 3.5 for two flat plates.

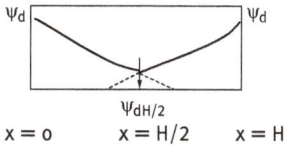

Fig. 3.5: Schematic representation of double-layer interaction for two flat plates.

The potential $\psi_{H/2}$ half way between the plates is no longer zero (as would be the case for isolated particles at $x \to \infty$). The potential distribution at an interdroplet distance H is schematically depicted by the full line in Fig. 3.5. The stern potential ψ_d is considered to be independent of the interdroplet distance distance. The dashed curves show the potential as a function of distance x to the Helmoltz plane, had the particles been at infinite distance.

For two spherical droplets of radius R and surface potential ψ_o and condition $\kappa R < 3$, the expression for the electrical double layer repulsive interaction is given by [6]:

$$G_{elec} = \frac{4\pi\varepsilon_r\varepsilon_o R^2 \psi_o^2 \exp-(\kappa h)}{2R + h},$$

(3.9)

where h is the closest distance of separation between the surfaces.

The above expression shows the exponential decay of G_{elec} with h. The higher the value of κ (i.e., the higher the electrolyte concentration), the steeper the decay, as schematically shown in Fig. 3.6. This means that at any given distance h, the double layer repulsion decreases with increase of electrolyte concentration.

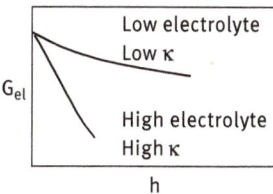

Fig. 3.6: Variation of G_{elec} with h at different electrolyte concentrations.

An important aspect of the double-layer repulsion is the situation during droplet approach. If at any stage the assumption is made that the double layers adjust to new conditions, so that equilibrium is always maintained, then the interaction takes place at constant potential. This would be the case if the relaxation time of the surface charge is much shorter than the time the particles are in each other's interaction sphere as a result of Brownian motion. However, if the relaxation time of the surface charge is appreciably longer than the time particles are in each other's interaction sphere, the charge rather than the potential will be the constant parameter. The constant charge leads to larger repulsion than the constant potential case.

Combination of G_{elec} and G_A results in the well-known theory of stability of colloids (DLVO Theory) [8, 9],

$$G_T = G_{elec} + G_A . \qquad (3.10)$$

A plot of G_T versus h is shown in Fig. 3.7, which represents the case at low electrolyte concentrations, i.e. strong electrostatic repulsion between the particles. G_{elec} decays exponentially with h, i.e. $G_{elec} \to 0$ as h becomes large. G_A is $\propto 1/h$, i.e. G_A does not decay to 0 at large h. At long distances of separation, $G_A > G_{elec}$, resulting in a shallow minimum (secondary minimum). At very short distances, $G_A \gg G_{elec}$, resulting in a deep primary minimum. At intermediate distances, $G_{elec} > G_A$, resulting in energy maximum, G_{max}, whose height depends on ψ_o (or ψ_d) and the electrolyte concentration and valency.

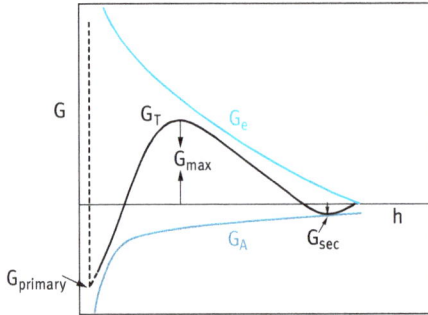

Fig. 3.7: Schematic representation of the variation of G_T with h according to the DLVO theory.

At low electrolyte concentrations ($< 10^{-2}$ mol dm^{-3} for a 1 : 1 electrolyte), G_{max} is high (> 25 kT), and this prevents particle aggregation into the primary minimum. The higher the electrolyte concentration (and the higher the valency of the ions), the lower the energy maximum. Under some conditions (depending on electrolyte concentration and particle size), flocculation into the secondary minimum may occur. This flocculation is weak and reversible. By increasing the electrolyte concentration, G_{max} decreases till at a given concentration it vanishes and particle coagulation occurs. This is illustrated in Fig. 3.8, which shows the variation of G_T with h at various electrolyte concentrations.

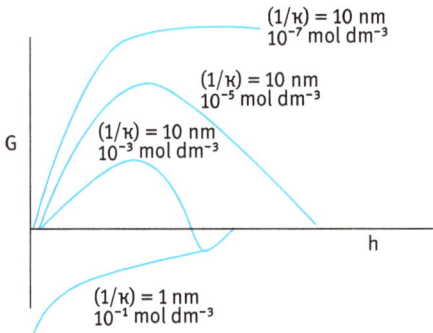

$(1/\kappa) = 10$ nm
10^{-7} mol dm^{-3}

$(1/\kappa) = 10$ nm
10^{-5} mol dm^{-3}

$(1/\kappa) = 10$ nm
10^{-3} mol dm^{-3}

G

h

$(1/\kappa) = 1$ nm
10^{-1} mol dm^{-3}

Fig. 3.8: Variation of G with h
at various electrolyte concentrations.

Since approximate formulae are available for G_{elec} and G_A, quantitative expressions for $G_T(h)$ can also be formulated. These can be used to derive expressions for the coagulation concentration, which is that concentration that causes every encounter between two emulsion droplets to lead to destabilization. Verwey and Overbeek [10] introduced the following criteria for transition between stability and instability:

$$G_T(= G_{elec} + G_A) = 0, \tag{3.11}$$

$$\frac{dG_T}{dh} = 0, \tag{3.12}$$

$$\frac{dG_{elec}}{dh} = -\frac{dG_A}{dh}. \tag{3.13}$$

Using the equations for G_{elec} and G_A, the critical coagulation concentration, c.c.c., could be calculated, as will be shown in Chapter 9. The theory predicts that the c.c.c. is directly proportional to the surface potential ψ_o and inversely proportional to the Hamaker constant A and the electrolyte valency Z. As will be shown in Chapter 9, the c.c.c is inversely proportional to Z^6 at high surface potential and inversely proportional to Z^2 at low surface potential.

3.3 Steric repulsion

This is produced by using non-ionic surfactants or polymers, e.g. alcohol ethoxylates, or A–B–A block copolymers PEO–PPO–PEO (where PEO refers to polyethylene oxide and PPO refers to polyropylene oxide), as is illustrated in Fig. 3.9.

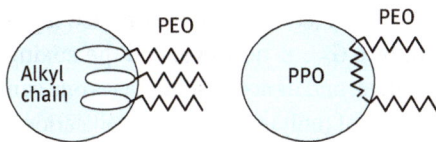

PEO

PEO

Alkyl
chain

PPO

Fig. 3.9: Schematic represetation
of adsorbed layers.

When two droplets, each with a radius R and containing an adsorbed polymer layer with a hydrodynamic thickness δ_h, approach each other at a surface–surface separation distance h that is smaller than $2\delta_h$, the polymer layers interact with each other, resulting in two main situations [11]: (i) the polymer chains may overlap; (ii) the polymer layer may undergo some compression.

In both cases, there will be an increase in the local segment density of the polymer chains in the interaction region. This is schematically illustrated in Fig. 3.10. The real situation is perhaps in between the above two cases, i.e. the polymer chains may undergo some interpenetration and some compression.

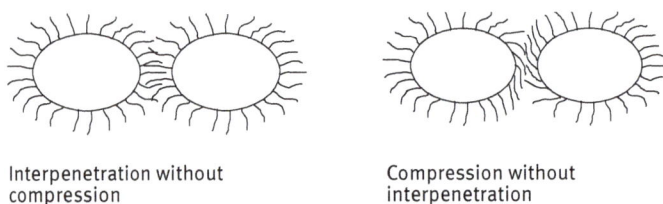

Interpenetration without compression

Compression without interpenetration

Fig. 3.10: Schematic representation of the interaction between droplets containing adsorbed polymer layers.

Provided the dangling chains (the A chains in A–B, A–B–A block or BA_n graft copolymers) are in a good solvent, this local increase in segment density in the interaction zone will result in strong repulsion as a result of two main effects [11]: (i) increase in the osmotic pressure in the overlap region as a result of the unfavourable mixing of the polymer chains, when these are in good solvent conditions – this is referred to as osmotic repulsion or mixing interaction and it is described by a free energy of interaction G_{mix}; (ii) reduction of the configurational entropy of the chains in the interaction zone – this entropy reduction results from the decrease in the volume available for the chains when these are either overlapped or compressed. This is referred to as volume restriction interaction, entropic or elastic interaction and it is described by a free energy of interaction G_{el}.

Combination of G_{mix} and G_{elec} is usually referred to as the steric interaction free energy, G_s, i.e.

$$G_s = G_{mix} + G_{el} . \tag{3.14}$$

The sign of G_{mix} depends on the solvency of the medium for the chains. If in a good solvent, i.e. the Flory–Huggins interaction parameter χ is less than 0.5, then G_{mix} is positive, and the mixing interaction leads to repulsion (see below). In contrast, if $\chi >$ 0.5 (i.e. the chains are in a poor solvent condition), G_{mix} is negative, and the mixing interaction becomes attractive. G_{el} is always positive, and hence in some cases one can produce stable dispersions in a relatively poor solvent (enhanced steric stabilisation).

3.3.1 Mixing interaction G_{mix}

This results from the unfavourable mixing of the polymer chains, when these are in a good solvent conditions. This is schematically shown in Fig. 3.11.

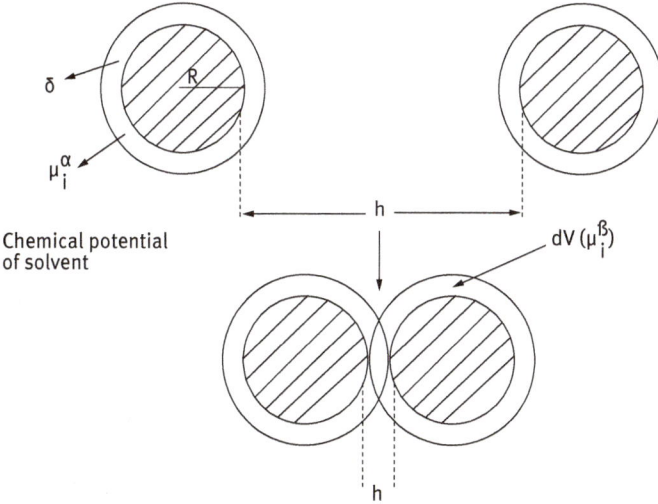

Fig. 3.11: Schematic representation of polymer layer overlap.

Consider two spherical droplets with the same radius and each containing an adsorbed polymer layer with thickness δ. Before overlap, one can define in each polymer layer a chemical potential for the solvent μ_i^α and a volume fraction for the polymer in the layer φ_2^α. In the overlap region (volume element dV), the chemical potential of the solvent is reduced to μ_i^β. This results from the increase in polymer segment concentration in this overlap region which is now φ_2^β.

In the overlap region, the chemical potential of the polymer chains is now higher than in the rest of the layer (with no overlap). This amounts to an increase in the osmotic pressure in the overlap region; as a result solvent will diffuse from the bulk to the overlap region, thus separating the particles, and hence a strong repulsive energy arises from this effect. The above repulsive energy can be calculated by considering the free energy of mixing of two polymer solutions, as for example treated by Flory and Krigbaum [12]. The free energy of mixing is given by two terms: (i) an entropy term that depends on the volume fraction of polymer and solvent, and (ii) an energy term that is determined by the Flory–Huggins interaction parameter:

$$\delta(G_{mix}) = kT(n_1 \ln \varphi_1 + n_2 \ln \varphi_2 + \chi n_1 \varphi_2), \tag{3.15}$$

where n_1 and n_2 are the number of moles of solvent and polymer with volume fractions φ_1 and φ_2, k is the Boltzmann constant and T is the absolute temperature.

The total change in free energy of mixing for the whole interaction zone, V, is obtained by summing over all the elements in V:

$$G_{mix} = \frac{2kTV_2^2}{V_1} v_2 \left(\frac{1}{2} - \chi \right) R_{mix}(h),$$
(3.16)

where V_1 and V_2 are the molar volumes of solvent and polymer respectively, v_2 is the number of chains per unit area and $R_{mix}(h)$ is geometric function which depends on the form of the segment density distribution of the chain normal to the surface, $\rho(z)$. k is the Boltzmann constant and T is the absolute temperature.

Using the above theory one can derive an expression for the free energy of mixing of two polymer layers (assuming a uniform segment density distribution in each layer) surrounding two spherical droplets as a function of the separation distance h between the particles [13].

The expression for G_{mix} is

$$G_{mix} = \left(\frac{2V_2^2}{V_1} \right) v_2 \left(\frac{1}{2} - \chi \right) \left(3R + 2\delta + \frac{h}{2} \right).$$
(3.17)

The sign of G_{mix} depends on the value of the Flory–Huggins interaction parameter χ: if $\chi < 0.5$, G_{mix} is positive and the interaction is repulsive; if $\chi > 0.5$, G_{mix} is negative and the interaction is attractive. The condition $\chi = 0.5$ and $G_{mix} = 0$ is termed the θ-condition. The latter corresponds to the case where the polymer mixing behaves as ideal, i.e. mixing of the chains does not lead to an increase or decrease of the free energy.

3.3.2 Elastic interaction G_{el}

This arises from the loss in configurational entropy of the chains on the approach of a second drop. As a result of this approach, the volume available for the chains becomes restricted, resulting in the loss of the number of configurations. This can be illustrated by considering a simple molecule, represented by a rod that rotates freely in a hemisphere across a surface (Fig. 3.12).

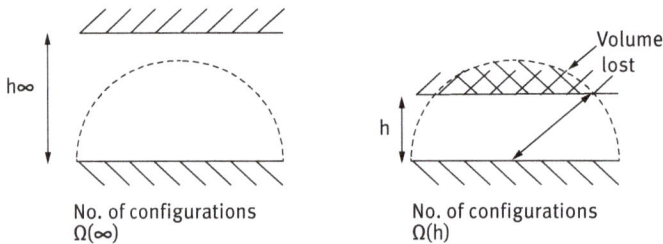

Fig. 3.12: Schematic representation of configurational entropy loss on approach of a second drop.

When the two surfaces are separated by an infinite distance ∞ the number of config-urations of the rod is $\Omega(\infty)$, which is proportional to the volume of the hemisphere. When a second drop approaches to a distance h such that it cuts the hemisphere (loos-ing some volume), the volume available to the chains is reduced, and the number of configurations become $\Omega(h)$, which is less than $\Omega(\infty)$. For two flat plates, G_{el} is given by the following expression:

$$\frac{G_{el}}{kT} = -2v_2 \ln\left[\frac{\Omega(h)}{\Omega(\infty)}\right] = -2v_2 R_{el}(h), \tag{3.18}$$

where $R_{el}(h)$ is a geometric function whose form depends on the segment density dis-tribution. It should be stressed that G_{el} is always positive and could play a major role in steric stabilisation. It becomes very strong when the separation distance between the particles becomes comparable to the adsorbed layer thickness δ.

3.3.3 Total energy of interaction

Combination of G_{mix} and G_{el} with G_A gives the total energy of interaction G_T (assuming there is no contribution from any residual electrostatic interaction), i.e.

$$G_T = G_{mix} + G_{el} + G_A. \tag{3.19}$$

A schematic representation of the variation of G_{mix}, G_{el}, G_A and G_T with surface–surface separation distance h is shown in Fig. 3.13.

G_{mix} increases very sharply with a decrease of h, when $h < 2\delta$. G_{el} increases very sharply with decrease of h, when $h < \delta$. G_T versus h shows a minimum, G_{min}, at sep-aration distances comparable to 2δ. When $h < 2\delta$, G_T shows a rapid increase with decrease in h. The depth of the minimum depends on the Hamaker constant A, the particle radius R and adsorbed layer thickness δ. G_{min} increases with increase of A and R. At a given A and R, G_{min} decreases with increase in δ (i.e. with increase of the

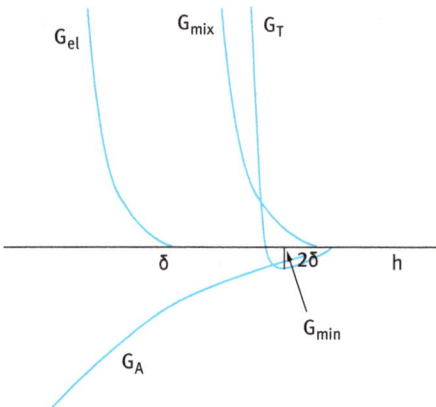

Fig. 3.13: Schematic representation of the energy–distance curve for a sterically stabilised emulsion.

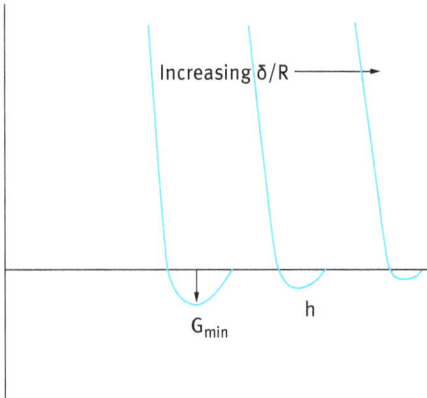

Fig. 3.14: Variation of G_T with h at various δ/R values.

molecular weight, M_w, of the stabiliser). This is illustrated in Fig. 3.14, which shows the energy–distance curves as a function of δ/R. The larger the value of δ/R, the smaller is the value of G_{min}. In this case the system may approach thermodynamic stability as is the case with nano-emulsions.

3.3.4 Criteria for effective steric stabilization

(i)   The droplets should be completely covered by the polymer (the amount of polymer should correspond to the plateau value). Any bare patches may cause flocculation, either by van der Waals attraction (between the bare patches) or by bridging flocculation (whereby a polymer molecule will become simultaneously adsorbed on two or more drops).

(ii) The polymer should be strongly "anchored" to the droplet surfaces to prevent any displacement during drop approach. This is particularly important for concentrated emulsions. For this purpose A–B, A–B–A block and BA_n graft copolymers are the most suitable, where the chain B is chosen to be highly insoluble in the medium and has a strong affinity to the surface or soluble in the oil. Examples of B groups for non-polar oils in aqueous media are polystyrene, polypropylene oxide and polymethylmethacrylate.

(iii) The stabilising chain A should be highly soluble in the medium and strongly solvated by its molecules. Examples of A chains in aqueous media are poly(ethylene oxide) and poly(vinyl alcohol).

(iv) δ should be sufficiently large ($> 5\,\text{nm}$) to prevent weak flocculation.

References

[1]   H. C. Hamaker, *Physica* (Utrecht) **4** (1937), 1058.

[2]   G. Gouy, *J. Phys.* 9 (1910), 457; *Ann. Phys.* **7**, 129 (1917).

[3]   D. L. Chapman, *Phil. Mag.* **25**, 475 (1913).

[4]   O. Stern, *Z. Electrochem.* **30** (1924), 508.

[5]   D. C. Grahame, *Chem. Rev.* **41** (1947), 44.

[6]   B. H. Bijesterbosch, Stability of Solid/Liquid Dispersions, in: Th. F. Tadros (ed.), *Solid/Liquid Dispersions*, Academic Press, London, 1987.

[7]   Tharwat Tadros, *Applied Surfactants* Wiley-VCH, Germany, 2005.

[8]   Tharwat Tadros, *Dispersion of Powders in Liquids and Stabilisation of Suspensions*, Wiley-VCH, Germany, 2013.

[9]   B. V. Deryaguin and L. Landua, *Acta Physicochem. USSR* **14** (1941), 633.

[10] E. J. W. Verwey and J. T. G. Overbeek, *Theory of Stability of Lyophobic Colloids*, Elsevier, Amsterdam, 1948.

[11] D. H. Napper, *Polymeric Stabilisation of Dispersions*, Academic Press, London, 1983.

[12] P. J. Flory and W. R. Krigbaum, *J. Chem. Phys.* **18** (1950), 1086.

[13] E. W., Fischer, *Kolloid Z.* **160** (1958), 120.

[14] E. L. Mackor and J. H. van der Waals, *J. Colloid Sci.* **7** (1951), 535.

[15] F. T. Hesselink, A. Vrij and J. T. G. Overbeek, *J. Phys. Chem.* **75** (1971), 2094.

4 Adsorption of surfactants at the oil/water interface

4.1 Introduction

As mentioned in Chapter 2, the Gibbs model [2] gives a definition of the interfacial tension as the change in free energy with area at constant temperature and composition of the interface (i.e. in absence of any adsorption):

$$\gamma = \left(\frac{\partial G^\sigma}{\partial A} \right)_{T,n_i}. \tag{4.1}$$

It is obvious from equation (4.1), that for a stable interface γ should be positive and the interfacial tension is the energy per unit area measured in $mJ\,m^{-2}$ or the force per unit length tangentially to the interface measured in units of $mN\,m^{-1}$.

There are generally two approaches for treating surfactant adsorption at the L/L interface. The first approach, adopted by Gibbs, treats adsorption as an equilibrium phenomenon, whereby the second law of thermodynamics may be applied using surface quantities. The second approach, referred to as the equation of state approach, treats the surfactant film as a two-dimensional layer with a surface pressure π that may be related the surface excess Γ (amount of surfactant adsorbed per unit area). Below, these two approaches are summarised.

4.2 The Gibbs adsorption isotherm

Gibbs [2] derived a thermodynamic relationship between the surface or interfacial tension γ and the surface excess Γ (adsorption per unit area). The starting point of this equation is the Gibbs–Deuhem equation given by equation (4.2). At equilibrium (where the rate of adsorption is equal to the rate of desorption), $dG^\sigma = 0$. At constant temperature, but in the presence of adsorption,

$$dG^\sigma = -S^\sigma \, dT + A \, d\gamma + \sum n_i \, d\mu_i \tag{4.2}$$

or

$$d\gamma = -\sum \frac{n_i^\sigma}{A} \, d\mu_i = -\sum \Gamma_i \, d\mu_i, \tag{4.3}$$

where $\Gamma_i = n_i^\sigma/A$ is the number of moles of component i and adsorbed per unit area.

Equation (4.3) is the general form for the Gibbs adsorption isotherm. The simplest case of this isotherm is a system of two components in which the solute (2) is the surface active component, i.e. it is adsorbed at the surface of the solvent (1). For such a case, equation (4.3) may be written as

$$-d\gamma = \Gamma_1^\sigma \, d\mu_1 + \Gamma_2^\sigma \, d\mu_2 \tag{4.4}$$

and if the Gibbs dividing surface is used, $\Gamma_1 = 0$ and

$$- d\gamma = \Gamma_{1,2}^\sigma \, d\mu_2 \tag{4.5}$$

where $\Gamma_{2,1}^\sigma$ is the relative adsorption of (2) with respect to (1). Since

$$\mu_2 = \mu_2^0 + RT \ln a_2^L, \tag{4.6}$$

or

$$d\mu_2 = RT \, d\ln a_2^L, \tag{4.7}$$

then

$$- d\gamma = \Gamma_{2,1}^\sigma RT \, d\ln a_2^L, \tag{4.8}$$

or

$$\Gamma_{2,1}^\sigma = -\frac{1}{RT} \left(\frac{d\gamma}{d\ln a_2^L} \right), \tag{4.9}$$

where a_2^L is the activity of the surfactant in bulk solution that is equal to $C_2 f_2$ or $x_2 f_2$, where C_2 is the concentration of the surfactant in mol dm^{-3} and x_2 is its mole fraction.

Equation (4.9) allows one to obtain the surface excess (abbreviated as Γ_2) from the variation of surface or interfacial tension with surfactant concentration. Note that $a_2 \sim C_2$, since in dilute solutions $f_2 \sim 1$. This approximation is valid since most surfactants have low c.m.c. (usually less than 10^{-3} mol dm^{-3}) and adsorption is complete at or just below the c.m.c.

The surface excess Γ_2 can be calculated from the linear portion of the γ–log C_2 curves before the c.m.c. Such γ–log C curves are illustrated in Fig. 4.1 for the air/water and O/W interfaces; [C_{SAA}] denotes the concentration of surface active agent in bulk solution. It can be seen that for the A/W interface, γ decreases from the value for water (72 mN m^{-1} at 20 °C) reaching about 25–30 mN m^{-1} near the c.m.c. This is clearly schematic since the actual values depend on the surfactant nature. For the O/W case, γ decreases from a value of about 50 mN m^{-1} (for a pure hydrocarbon-water interface) to ~ 1–5 mN m^{-1} near the c.m.c. (again depending on the nature of the surfactant).

As mentioned above, Γ_2 can be calculated from the slope of the linear position of the curves shown in Fig. 4.1 just before the c.m.c. is reached. From Γ_2, the area per surfactant ion or molecule can be calculated since

$$\text{Area/molecule} = \frac{1}{\Gamma_2 N_{av}} \tag{4.10}$$

where N_{av} is the Avogadro constant. Determining the area per surfactant molecule is very useful, since it gives information on surfactant orientation at the interface. For example, for ionic surfactants such as alkyl sulphates, the area per surfactant is determined by the area occupied by the alkyl chain and head group if these molecules lie flat at the interface. In this case the area per molecule increases with an increase in the alkyl chain length. For vertical orientation, the area per surfactant ion is determined by whatever is occupied by the charged head group, which at low electrolyte

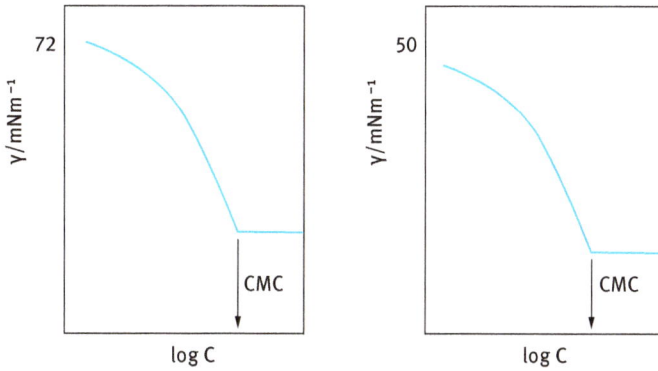

Fig. 4.1: Variation of surface and interfacial tension with log [C_{SAA}] at the air-water and oil-water interface.

concentration will be in the region of 0.40 nm^2. Such an area is larger than the geometrical area occupied by a sulphate group, as a result of the lateral repulsion between the head group. When electrolytes are added, this lateral repulsion is reduced and the area/surfactant ion for vertical orientation will be lower than 0.4 nm^2 (reaching in some cases 0.2 nm^2).

Another important point can be made from the γ–log C curves. At concentration just before the break point, one has the condition of constant slope, which indicates that saturation adsorption has been reached:

$$\left(\frac{\partial \gamma}{\partial \ln a_2} \right)_{p,T} = \text{const.} \tag{4.11}$$

Just above the break point,

$$\left(\frac{\partial \gamma}{\partial \ln a_2} \right)_{p,T} = 0, \tag{4.12}$$

indicating the constancy of γ with log C above the c.m.c. Integration of equation (4.12) gives

$$\gamma = \text{const.} \cdot \ln a_2 \tag{4.13}$$

Since γ is constant in this region, then a_2 must remain constant. This means that the addition of surfactant molecules above the c.m.c. must result in an association to form units (micellar) with low activity.

The hydrophilic head group may be unionised, e.g. alcohols or poly(ethylene oxide) alkane or alkyl phenol compounds, weakly ionised such as carboxylic acids or strongly ionised such as sulphates, sulphonates and quaternary ammonium salts. The adsorption of these different surfactants at the air/water and oil/water interface depends on the nature of the head group. With non-ionic surfactants, repulsion between the head groups is small, and these surfactants are usually strongly adsorbed at the surface of water from very dilute solutions. Non-ionic surfactants have much lower c.m.c. values compared to ionic surfactants with the same alkyl chain length. Typically, the c.m.c. is in the region of 10^{-5}–10^{-4} mol dm^{-3}. Such non-ionic surfactants form closely packed adsorbed layers at concentrations lower than their c.m.c. values. The activity coefficient of such surfactants is close to unity and is only slightly affected by addition of moderate amounts of electrolytes (or a change in the pH of the solution). Thus, non-ionic surfactant adsorption is the simplest case, since the solutions can be represented by a two component system and the adsorption can be accurately calculated using equation (4.9).

With ionic surfactants, on the other hand, the adsorption process is relatively more complicated, since one has to consider the repulsion between the head groups and the effect of presence of any indifferent electrolyte. Moreover, the Gibbs adsorption equation has to be solved taking into account the surfactant ions, the counter-ion and any indifferent electrolyte ions present. For a strong surfactant electrolyte such as an Na^+R^-,

$$\Gamma_2 = \frac{1}{2RT} \frac{\partial \gamma}{\partial \ln a\pm} \tag{4.14}$$

The factor of 2 in equation (4.14) arises because both the surfactant ion and the counter-ion must be adsorbed to maintain neutrally, and $d\gamma/d\ln a\pm$ is twice as large as for an un-ionised surfactant.

If a non-adsorbed electrolyte such as NaCl is present in large excess, then any increase in concentration of Na^+R^- produces a negligible increase in Na^+ ion concentration, and therefore $d\mu_{Na}$ becomes negligible. Moreover, $d\mu_{Cl}$ is also negligible, so that the Gibbs adsorption equation reduces to

$$\Gamma_2 = -\frac{1}{RT} \left(\frac{\partial \gamma}{\partial \ln C_{NaR}} \right) \tag{4.15}$$

i.e. it becomes identical to that for a non-ionic surfactant.

The above discussion, clearly illustrates that for calculation of Γ_2 from the γ–log C curve one has to consider the nature of the surfactant and the composition of the medium. For non-ionic surfactants the Gibbs adsorption equation (4.9) can be directly used. For ionic surfactant, in absence of electrolytes the right hand side of equation (4.9) should be divided by 2 to account for surfactant dissociation. This factor disappears in the presence of a high concentration of an indifferent electrolyte.

4.3 Equation of state ppproach

In this approach, one relates the surface pressure π with the surface excess Γ_2. The surface pressure is defined by the equation [3–5]

$$\pi = \gamma_0 - \gamma , \tag{4.16}$$

where γ_o is the surface or interfacial tension before adsorption and γ that after adsorption.

For an ideal surface film, behaving as a two-dimensional gas the surface pressure π is related to the surface excess Γ_2 by the equation

$$\pi A = n_2 RT \tag{4.17}$$

or

$$\pi = (n_2/A)RT = \Gamma_2 RT. \tag{4.18}$$

Differentiating equation (4.16) at constant temperature,

$$d\pi = RT \, d\Gamma_2. \tag{4.19}$$

Using the Gibbs equation,

$$d\pi = -d\gamma = \Gamma_2 RT \, d\ln a_2 \approx \Gamma_2 RT \, d\ln C_2. \tag{4.20}$$

Combining equations (4.19) and (4.20),

$$d\ln \Gamma_2 = d\ln C_2 \tag{4.21}$$

or

$$\Gamma_2 = KC_2^{\alpha} \tag{4.22}$$

Equation (4.22) is referred to as the Henry's law isotherm, which predicts a linear relationship between Γ_2 and C_2.

It is clear that equations (4.16) and (4.19) are based on an idealised model in which the lateral interaction between the molecules has not been considered. Moreover, in this model the molecules are considered to be dimensionless. This model can only be applied at very low surface coverage, where the surfactant molecules are so far apart that lateral interaction may be neglected. Moreover, under these conditions the total area occupied by the surfactant molecules is relatively small compared to the total interfacial area.

At significant surface coverage, the above equations have to be modified to take into account both lateral interaction between the molecules as well as the area occupied by them. Lateral interaction may reduce π if there is attraction between the chains (e.g. with most non-ionic surfactant), or it may increase π as a result of repulsion between the head groups in the case of ionic surfactants.

Various equation of state have been proposed, taking into account the above two effects, in order to fit the π–A data. The two-dimensional van der Waals equation of state is probably the most convenient for fitting these adsorption isotherms, i.e.

$$\left(\pi + \frac{(n_2)^2\alpha}{A_2}\right)(A - n_2 A_2^0) = n_2 RT, \tag{4.23}$$

where A_2^o is the excluded or co-area of type 2 molecule in the interface and α is a parameter which allows for lateral interaction.

Equation (4.22) leads to the following theoretical adsorption isotherm, using the Gibbs equation:

$$C_2^\alpha = K_1 \left(\frac{\theta}{1-\theta} \right) \exp \left(\frac{\theta}{1-\theta} - \frac{2\alpha\theta}{a_2^o RT} \right),$$ (4.24)

where θ is the surface coverage ($\theta = \Gamma_2 / \Gamma_{2,\max}$), K_1 is constant that is related to the free energy of adsorption of surfactant molecules at the interface ($K_1 \propto \exp\left(-\Delta G_{ads}/kT\right)$) and a_2^o is the area/molecule.

For a charged surfactant layer, equation (4.21) has to be modified to take into account the electrical contribution from the ionic head groups, i.e.

$$C_2^\alpha = K_1 \left(\frac{\theta}{1-\theta} \right) \exp \left(\frac{\theta}{1-\theta} \right) \exp \left(\frac{e\psi_o}{kT} \right),$$ (4.25)

where ψ_o is the surface potential. Equation (4.25) shows how the electrical potential energy (ψ_o/kT) of adsorbed surfactant ions affects the surface excess. Assuming that the bulk concentration remains constant, then ψ_o increase as θ increases. This means that $[\theta/(1-\theta)]\exp[\theta/(1-\theta)]$ increases less rapidly with C_2, i.e. adsorption is inhibited as a result of ionisation.

4.4 The Langmuir, Szyszkowski and Frumkin equations

In addition to the Gibbs equation, three other equations have been suggested which relate the surface excess Γ_1, surface or interfacial tension, and equilibrium concentration in the liquid phase C_1. The Langmuir equation [6] relates Γ_1 to C_1 by

$$\Gamma_1 = \frac{\Gamma_m C_1}{C_1 + a},$$ (4.26)

where Γ_m is the saturation adsorption at monolayer coverage by surfactant molecules and a is a constant related to the free energy of adsorption ΔG_{ads}^o,

$$a = 55.3 \exp \left(\frac{\Delta G_{ads}^o}{RT} \right),$$ (4.27)

where R is the gas constant and T is the absolute temperature.

A linear form of the Gibbs equation is

$$\frac{1}{\Gamma_1} = \frac{1}{\Gamma_m} + \frac{a}{\Gamma_m C_1}.$$ (4.28)

Equation (4.28) shows that a plot of $1/\Gamma_1$ versus $1/C_1$ gives a straight line from which Γ_m and a can be calculated from the intercept and slope of the line.

The Szyszkowski [7] equation gives a relationship between the surface pressure π and bulk surfactant concentration C_1; it is a form of equation of state:

$$\gamma_0 - \gamma = \pi = 2.303 RT\, \Gamma_m \log\left(\frac{C_1}{a} + 1\right). \tag{4.29}$$

The Frumkin equation [8] is another equation of state:

$$\gamma_0 - \gamma = \pi = -2.303 RT\, \Gamma_m \log\left(1 - \frac{\Gamma_1}{\Gamma_m}\right). \tag{4.30}$$

4.5 Effectiveness of surfactant adsorption at the liquid/liquid interface

The surface excess concentration at surface saturation, Γ_m is a useful measure of the effectiveness of the surfactant at the liquid/liquid interface [9]. This is an important factor in determining such properties of the surfactant in several processes such as emulsification and emulsion stability. In most cases, a tightly packed, coherent film obtained by vertically oriented surfactant molecules is required. The area occupied by a surfactant molecule consisting of a linear alkyl chain and one hydrophilic group, either ionic or non-ionic, is larger than the cross-sectional area of an aliphatic chain (which is $0.2\,\text{nm}^2$), indicating that the area occupied by a surfactant molecule is de-termined by the area occupied by the hydrated hydrophilic chain. For example for a series of alkyl sulphates the area occupied by the surfactant molecule is in the region of $0.5\,\text{nm}^2$, which is the cross-sectional area of a sulphate group. As mentioned before, addition of electrolytes reduces the area occupied by the surfactant molecule due to charge screening of the sulphate group. With non-ionic surfactants based on a polyethylene oxide (PEO) hydrophilic group the amount of adsorption at saturation Γ_m decreases with an increase of the PEO chain length, and this results in an increase in the area occupied by a singly surfactant molecule.

4.6 Efficiency of adsorption of surfactant at the liquid/liquid interface

The efficiency of surfactant adsorption at the liquid/liquid interface can be determined by measuring the surfactant concentration that produces a given amount of adsorption at the interface [9]. This can also be related to the free-energy change involved in the adsorption. A convenient measure of the efficiency of adsorption is the negative logarithm of surfactant concentration C in bulk solution required to produce a $20\,\text{mN}\,\text{m}^{-1}$ reduction in the interfacial tension γ:

$$-\log C_{(-\Delta\gamma=20)} \equiv pC_{20}. \tag{4.31}$$

Observation of various γ–log C results for various surfactants at the oil/water interface shows that when γ is reduced by $20\,\mathrm{mN\,m^{-1}}$, the surface excess concentration Γ_1 is close to its saturation value Γ_m. This is confirmed by the use of the Frumkin equation (4.29). For many surfactant systems Γ_m is in the range 1–$4.4 \cdot 10^{-6}\,\mathrm{mol\,m^{-2}}$. Solving for Γ_1 in the Frumkin equation when $\pi = \gamma_0 - \gamma = 20\,\mathrm{mN\,m^{-1}}$, $\Gamma_m = 1$–$4.4 \cdot 10^{-6}\,\mathrm{mol\,m^{-2}}$, $\Gamma_1 = 0.84$–$0.99\Gamma_m$ at $25\,^\circ\mathrm{C}$, indicating that when γ is reduced by $20\,\mathrm{mN\,m^{-1}}$, the surface concentration is 84–$99\,\%$ saturated.

p20 can be related to the free energy change on adsorption at infinite dilution ΔG° by application of the Langmuir [6] and Szyszkowski [7] equations (4.25) and (4.28). As mentioned above for $\pi = 20\,\mathrm{mN\,m^{-1}}$, $\Gamma_1 = 0.84$–$0.99\Gamma_m$. From the Langmuir equation $C_1 = 5.2$–$999xa$. Thus,

$$\log\left[\left(\frac{C_1}{a}\right) + 1\right] \approx \log\left(\frac{C_1}{a}\right). \tag{4.32}$$

The Szyszkowski equation becomes

$$\pi = \gamma_0 - \gamma = -2.303RT\,\Gamma_m \log\left(\frac{C_1}{a}\right), \tag{4.33}$$

and

$$\log\left(\frac{1}{C_1}\right)_{\pi=20} = -\left(\log a + \frac{\gamma_0 - \gamma}{2.303RT\,\Gamma_m}\right). \tag{4.34}$$

since

$$a = 55\exp\left(\frac{\Delta G^\circ}{RT}\right), \tag{4.35}$$

$$\log a = 1.74 + \frac{\Delta G^\circ}{2.303RT}, \tag{4.36}$$

$$\log\left(\frac{1}{C_1}\right)_{\pi=20} \equiv pC_{20} \equiv -\left(\frac{\Delta G_o}{2.303RT} + 1.74 + \frac{20}{2.303RT\,\Gamma_m}\right). \tag{4.37}$$

For a straight-chain surfactant molecule consisting of m $-CH_2-$ groups and one hydrophilic head group (h), ΔG° can be broken into the standard free energy associated with the transfer of the terminal CH_3, the $-CH_2-$ groups of the hydrocarbon chain and the hydrophilic group h from the interior of the liquid phase to the interface at $\pi = 20\,\mathrm{mN\,m^{-1}}$:

$$\Delta G^\circ = m\,\Delta G^\circ(-CH_2-) + \Delta G^\circ(h) + \text{const.} \tag{4.38}$$

For a homologous series of surfactants with the same hydrophilic group h, the value of Γ_m and the area per surfactant molecule does not change much with an increase in the number of C atoms, and $\Delta G^\circ(h)$ can be considered constant. In this case pC_{20} can be related to the free-energy change per $-CH_2-$ group by

$$pC_{20} = \left[\frac{-\Delta G^\circ(-CH_2-)}{2.303RT}\right]m + \text{const.} \tag{4.39}$$

Equation (4.39) shows that pC_{20} is a linear function of the number of C atoms in the surfactant chain m. The larger the value of pC_{20} the more efficiently the surfactant is adsorbed at the interface and the more efficiently it reduces the interfacial tension.

The efficiency of surfactant adsorption at the liquid/liquid interface as measured by pC_{20} increases with an increase in the number of C atoms in the surfactant. Straight alkyl chains are generally more efficient than branched ones with the same C number. A single hydrophilic group at the end of the hydrophobic chain gives more efficient adsorption than where the hydrophilic group is located at the centre of the chain. Non-ionic and zwitter-ionic surfactants generally give more efficient adsorption compared to ionic ones. With ionic surfactants, increase of the ionic strength of the aqueous solution increases the efficiency of surfactant adsorption.

4.7 Adsorption from mixtures of two surfactants

Mixtures of two or more different types of surfactants often show a "synergistic" interaction [10, 11]. In other words, the interfacial properties of the mixture are more pronounced than those of the individual components themselves. These mixtures are widely used in many industrial applications such as in emulsification and emulsion stability. A study of the adsorption of the individual components in the mixture and the interaction between them allows us to understand the role played by each component. This will also enables us to make the right selection of surfactant mixtures for a specific application.

The Gibbs adsorption equation (4.9) for two surfactants in dilute solution can be written as

$$d\gamma = RT(\Gamma_1 \, d\ln a_1 + \Gamma_2 RT \, d\ln a_2), \tag{4.40}$$

where Γ_1 and Γ_2 are the surface excess concentrations of the two surfactants at the interface, and a_1 and a_2 are their respective activities in solution. Since the solutions are dilute, a_1 and a_2 can be replaced by the molar concentrations C_1 and C_2.

Using equation (4.10),

$$\Gamma_1 = \frac{1}{RT}\left(\frac{-\partial\gamma}{\partial\ln C_1}\right)_{C_2} = \frac{1}{2.303RT}\left(\frac{-\partial\gamma}{\partial\log C_1}\right)_{C_2}, \tag{4.41}$$

$$\Gamma_2 = \frac{1}{RT}\left(\frac{-\partial\gamma}{\partial\ln C_2}\right)_{C_1} = \frac{1}{2.303RT}\left(\frac{-\partial\gamma}{\partial\log C_2}\right)_{C_1}. \tag{4.42}$$

Therefore, the concentration of each surfactant at the interface can be calculated from the slope of the γ–log C plot of each surfactant, holding the solution concentration of the second surfactant constant.

For ideal mixing of two surfactants (with no net interaction), C_1 and C_2 are given by the expressions [12]

$$C_1 = C_1^o f_1 X_1 , \tag{4.43}$$

$$C_2 = C_2^o f_2 X_2 , \tag{4.44}$$

where f_1 and f_2 are the activity coefficients of surfactant 1 and 2, respectively, X_1 is the mole fraction of surfactant 1 at the interface, i.e. $X_1 = 1 - X_2$, C_1^o is the molar concentration required to attain a given interfacial tension in a solution of pure surfactant 1, and C_2^o is the molar concentration required to attain a given interfacial tension in a solution of pure surfactant 2.

For non-ideal mixing, i.e. when there is interaction between the surfactant molecules, the activity coefficient has to include the interaction parameter β^σ:

$$\ln f_1 = \beta^\sigma (1 - X_1)^2 , \tag{4.45}$$

$$\ln f_2 = \beta^\sigma (X_1)^2 . \tag{4.46}$$

Combining equations (4.43)–(4.45),

$$\frac{(X_1)^2 \ln (C_1/C_1^o X_1)}{(1 - X_1)^2 \ln \left[\frac{C_2}{C_2^o (1-X_1)} \right]} = 1 . \tag{4.47}$$

The interfacial tension-total surfactant concentration (C_t) curves for two pure surfactants, and their mixture at a fixed value α, the mole fraction of surfactant 1 in the total surfactant in the solution phase, are used (Fig. 4.2) to determine C_1 ($= \alpha C_{12}$), C_1^o, C_2 [$= (1 - \alpha)C_{12}$] and C_2^o, the molar concentration at the same surface tension. Equation (4.46) permits it to be solved iteratively for X_1 and X_2 ($= 1 - X_1$). The ratio of surfactant 1 : surfactant 2 at the interface at that particular value of α is then X_1/X_2.

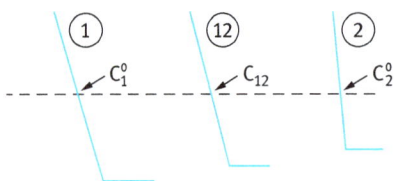

Fig. 4.2: Evaluation of X_1 and X_2: (1) pure surfactant 1; (2) pure surfactant 2; (12) Mixture of 1 and 2 at a fixed value of α.

4.8 Adsorption of macromolecules

Most theories and experimental results on the adsorption of macromolecules have been devoted to the solid/liquid interface. At such an interface, the polymer is generally thought to adsorb in a loop-train type of conformation, with the tails at the end of the molecule [13]. Segments in trains are in contact with the solid surface, while those in loops or tails are immersed in the liquid phase. At a liquid/liquid interface,

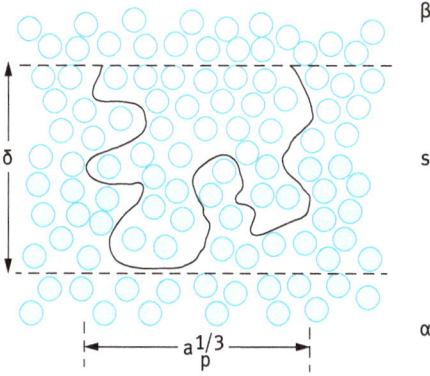

Fig. 4.3: Typical conformation of a homopolymer chain at a liquid/liquid interface (α/β).

segments of a given homopolymer will partition themselves between the two liquid phases, in a manner which reflects the two segment-solvent interaction parameters ε_{13} and ε_{23} (where 1 and 2 refer to the two liquids and 3 to the polymer segments). A typical situation for an adsorbed homopolymer is illustrated in Fig. 4.3.

The time-average conformation adopted will be governed by the net energy-entropy balance for the polymer at the interface. The interfacial region with thickness δ is the region depicted by s in Fig. 4.3. There are two major distinctions to be made when comparing the adsorption of macromolecules with that of small molecules. First, we have to distinguish between the adsorption of the whole molecule Γ_p^s and that of segments (designated 3) Γ_3^s. Second, because the adsorption of macromolecules is irreversible [13], one cannot apply the second law of thermodynamics discussed above.

Γ_p^s is related to the concentration of polymer in the two phases C_p^α and C_p^β by

$$\Gamma_p^s = \frac{n_p^s}{\Delta A} = \frac{n_p - (C_p^\alpha V^\alpha + C_p^\beta V^\beta)}{\Delta A} . \tag{4.48}$$

The effective area per polymer molecule a_p is simply given by

$$a_p = \frac{1}{\Gamma_p^s} . \tag{4.49}$$

Determination of Γ_p^s as a function of C_p^α or C_p^β is one of the major considerations in any study of polymer adsorption at the liquid/liquid interface. Establishment of polymer adsorption isotherms is more complex for the liquid/liquid interface than for the solid/liquid interface. In the first place, both the equilibrium values C_p^α and C_p^β are required, and in the second place ΔA has to be known. If an emulsion is used to achieve a high value of ΔA, the latter is only determined after the emulsion has been formed in the presence of polymer. This means that one can effectively determine Γ_p^s at maximum coverage concentration. Establishment of the low coverage value can only be carried out using a planar liquid/liquid interface.

When using block copolymers of the A–B or A–B–A types and graft copolymers of the BA_n type (where B is the hydrophobic chain and A is the hydrophilic chain), then in an O/W emulsion the lipophilic B chain will reside in the oil phase, whereas the hydrophilic A chain(s) will reside in the aqueous phase. In this case one needs to know the interaction between the B chain and the oil and the interaction between the A chain(s) and the aqueous phase. A useful method is to apply the solubility parameter concept as will be discussed in Chapter 7.

4.9 Interfacial tension measurements

These methods may be classified into two categories: (i) those in which the properties of the meniscus is measured at equilibrium, e.g., pendent drop or sessile drop profile and Wilhelmy plate methods and (ii) those where the measurement is made under non- or quasi-equilibrium conditions such as the drop volume (weight) or the du Noüy ring method. The latter methods are faster, although they have the disadvantage of premature rupture and expansion of the interface, causing adsorption depletion. They are also inconvenient for measurement of the interfacial tension in the presence of macromolecular species, where adsorption is slow. This problem is overcome with the equilibrium (static) methods. For measurement of low interfacial tensions ($< 0.1\,\mathrm{mN\,m^{-1}}$) the spinning drop technique is applied. Below a brief description of each of these techniques is given.

4.9.1 The Wilhelmy plate method

In this method [14] a thin plate made from glass (e.g., a microscope cover slide) or platinum foil is either detached from the interface (non-equilibrium condition) or is weight measured statically using an accurate microbalance. In the detachment method, the total force F is given by the weight of the plate W, and the interfacial tension force

$$F = W + \gamma p, \tag{4.50}$$

where p is the "contact length" of the plate with the liquid, i.e., the plate perimeter. Provided the contact angle of the liquid is zero, no correction is required for equation (4.48). Thus, the Wilhelmy plate method can be applied in the same manner as the du Noüy technique described below.

The static technique may be applied for following the interfacial tension as a function of time (to follow the kinetics of adsorption) until equilibrium is reached. In this case, the plate (roughened to ensure a zero contact angle) is suspended from one arm of a microbalance and allowed to penetrate the upper liquid layer (usually the oil) into the aqueous phase to ensure wetting of the plate. The whole vessel is then lowered to bring the plate in the oil phase. At this point the microbalance is adjusted to counteract the weight of the plate (i.e., its weight now becomes zero). The vessel is then raised until the plate touches the interface. The increase in weight ΔW is given by the following equation:

$$\Delta W = \gamma p \cos \theta , \tag{4.51}$$

where θ is the contact angle. If the plate is completely wetted by the lower liquid as it penetrates, $\theta = 0$ and γ may be calculated directly from ΔW. Care should always be taken that the plate is completely wetted by the aqueous solution. For that purpose, a roughened platinum or glass plate is used to ensure a zero contact angle. However, if the oil is more dense than water, a hydrophobic plate is used so that when the plate penetrates through the upper aqueous layer and touches the interface it is completely wetted by the oil phase.

4.9.2 The pendent drop method

If a drop of oil is allowed to hang from the end of a capillary that is immersed in the aqueous phase, it will adopt an equilibrium profile, as shown in Fig. 4.4, that is a unique function of the tube radius, the interfacial tension, its density and the gravitational field.

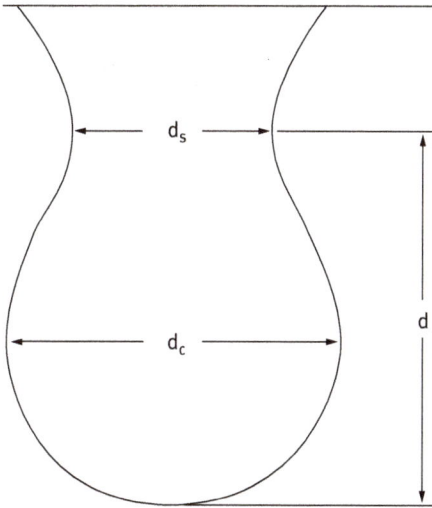

Fig. 4.4: Schematic representation of the profile of a pendent drop.

The shape of the bounded menisci (those with only one solid surface supporting the liquid forming the meniscus, as is the case with the pendent drop) can be described by a single parameter β, derived by Bashforth and Adams [15]:

$$\beta = \frac{\Delta \rho g b^2}{\gamma} , \tag{4.52}$$

where $\Delta\rho$ is the density difference between the two phases, g is the gravity force and b is the radius of a drop at its apex.

Andreas et al. [16] suggested that the most conveniently measurable shape dependent quantity is S:

$$S = \frac{d_s}{d_e} , \tag{4.53}$$

where d_e is the equatorial diameter and d_s is the diameter measured at a distance d from the bottom of the drop. The difficulty in measuring b is overcome by combining it with β and defining a new quantity H, given by

$$H = -\beta \left(\frac{d_e}{b}\right)^2 . \tag{4.54}$$

The interfacial tension is given by the following equation:

$$\gamma = \frac{\Delta\rho g d_e^2}{H} , \tag{4.55}$$

The relationship between H and the experimental values of d_s/d_e has been obtained empirically using pendent drops of water. Accurate values of H have been obtained by Niederhauser and Bartell [17].

The pendent drop technique is used for accurate measurements of interfacial tension ($\pm 0.1\%$), in particular with the development of image analysis methods for obtaining the drop shape. The technique can be applied following the kinetics of surfactant adsorption, and in particular for the slow process of macromolecular adsorption.

4.9.3 Sessile drop method

This is similar to the pendent drop method, except in this case a drop of the liquid with the higher density is placed on a flat plate immersed in the second liquid, as schematically shown in Fig. 4.5.

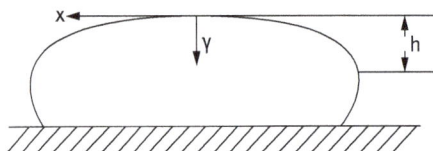

Fig. 4.5: Schematic representation of a sessile drop.

The interfacial tension γ is calculated by

$$\gamma = \frac{\Delta\rho g b^2}{\beta} , \tag{4.56}$$

while β is difficult to determine, the Bashforth–Adam tables [15] give x_e/b as a function of β, where x_e is the equatorial radius, so that equation (4.56) may be written in the form

$$\gamma = \Delta\rho \frac{gx_e^2}{[f(\beta)]^2} , \qquad (4.57)$$

with $f(\beta) = x_e/b$.

Since x_e can be determined accurately, the experimental problem reduces to the determination of β. This is achieved by comparing the drop profile to a theoretical set of profiles, for various values of β as given by the tables.

The sessile drop method has the same accuracy ($\pm 0.1\,\%$) as the pendent drop technique, and it allows one to measure the interfacial tension as a function of time (i.e. measuring the kinetics of surfactant adsorption).

4.9.4 The Du Nouy ring method

Basically one measures the force required to detach a ring or loop of wire with radius R from the liquid/liquid interface [18]. As a first approximation, the detachment force is taken to be equal to the interfacial tension γ multiplied by the perimeter of the ring. The total force F in detaching the ring from the interface is the sum of its weight w and the interfacial force:

$$F = W + 4\pi R\gamma . \qquad (4.58)$$

Harkins and Jordan [19] introduced a correction factor f (that is a function of meniscus volume V and radius r of the wire) for more accurate calculation of γ from F, i.e.

$$f = \frac{\gamma}{\gamma_{ideal}} = f\left(\frac{R^3}{V}, \frac{R}{r}\right) . \qquad (4.59)$$

Values of the correction factor f were tabulated by Harkins and Jordan [19]. Some theoretical account of f was given by Freud and Freud [20].

When using the du Noüy method for measurement of γ one must be sure that the ring is kept horizontal during the measurement. Moreover, the ring should be free from any contaminant, which is usually achieved by using a platinum ring that is flamed before use.

4.9.5 The drop volume (weight) method

Here one determines the volume V (or weight W) of a drop of liquid (immersed in the second less dense liquid) which becomes detached from a vertically mounted capillary tip having a circular cross section of radius r. The ideal drop weight W_{ideal} is given by the expression

$$W_{ideal} = 2\pi r\gamma . \qquad (4.60)$$

In practice, a weight W is obtained which is less than W_{ideal} because a portion of the drop remains attached to the tube tip. Thus, equation (4.60) should include a correction factor φ which is a function of the tube radius r and some linear dimension of the drop, i.e. $V^{1/3}$. Thus,

$$W = 2\pi r \gamma \varphi \left(\frac{r}{V^{1/3}} \right). \tag{4.61}$$

Values of $(r/V^{1/3})$ have been tabulated by Harkins and Brown [21]. Lando and Oakley [22] used a quadratic equation to fit the correction function to $(r/V^{1/3})$. A better fit was provided by Wlkinson and Kidwell [23].

4.9.6 The spinning drop method

This method is particularly useful for the measurement of very low interfacial tensions ($< 10^{-1}$ mN m^{-1}), which are particularly important in applications such as spontaneous emulsification and the formation of microemulsions. Such low interfacial tensions may also be reached with emulsions, particularly when mixed surfactant films are used. A drop of the less dense liquid A is suspended in a tube containing the second liquid B. On rotating the whole mass (Fig. 4.6) the drop of the liquid moves to the centre. With increasing speed of revolution, the drop elongates as the centrifugal force opposes the interfacial tension force that tends to maintain the spherical shape, i.e. having a minimum surface area.

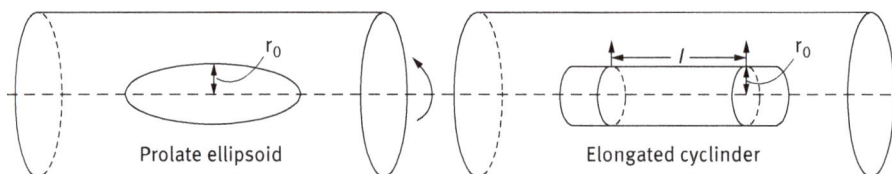

Fig. 4.6: Schematic representation of a spinning drop: (a) prolate ellipsoid; (b) elongated cylinder.

An equilibrium shape is reached at any given speed of rotation. At moderate speeds of rotation, the drop approximates to a prolate ellipsoid, whereas at very high speeds of revolution, the drop approximates to an elongated cylinder. This is schematically shown in Fig. 4.6.

When the shape of the drop approximates a cylinder (Fig. 4.4 (b)), the interfacial tension is given by the following expression [24]:

$$\gamma = \frac{\omega_2 \Delta \rho r_0^4}{4}, \tag{4.62}$$

where ω is the speed of rotation, $\Delta\rho$ is the density difference between the two liquids A and B and r_0 is the radius of the elongated cylinder. Equation (4.62) is valid when the length of the elongated cylinder is much larger than r_0.

References

[1] E. A. Guggenheim, *Thermodynamics*, 5th ed., p. 45, North Holland, Amsterdam, 1967.
[2] J. W. Gibbs, *Collected Works*, vol. 1, p. 219, Longman, New York, 1928.
[3] R. Aveyard and B. J. Briscoe, *Trans. Faraday Soc.* **66** (1970), 2911.
[4] R. Aveyard, B. J. Briscoe and J. Chapman, *Trans. Faraday Soc.* **68** (1972), 10.
[5] R. Aveyard and J. Chapman, *Can. J. Chem.* **53** (1975), 916.
[6] I. Langmuir, *J. Amer. Chem. Soc.* **39** (1917), 1848.
[7] B. Szyszkowski, *Z. Phys. Chem.* **64** (1908), 385.
[8] A. Frumkin, *Z. Phys. Chem.* **116** (1925), 466.
[9] M. J. Rosen and J. T. Kunjappu, *Surfactants and Interfacial Phenomena*, John Wiley and Sons, New Jersey, 2012.
[10] R. de Lisi, A. Inglese, S. Milioto and A. Pellento, *Langmuir* **13** (1997), 192.
[11] T. Y. Nakano, G. Sugihara, T. Nakashima and S. C. Yu, *Langmuir* **18** (2002), 8777.
[12] M. J. Rosen and X. Y. Hua, *J. Colloid Interface Sci.* **86** (1982), 164.
[13] 31. G. J. Fleer,.M. A. Cohen-Stuart, J. M. H. M. Scheutjens, T. Cosgrove and B. Vincent, *Polymers at Interfaces*, Chapman and Hall, London, 1993.
[14] L. Wilhelmy, *Ann. Phys.* **119** (1863), 177.
[15] F. Bashforth and J. C. Adams, *An Attempt to Test the Theories of Capillary Action*, University Press, Cambridge, 1883.
[16] J. M. Andreas, E. A. Hauser and W. B. Tucker, *J. Phys. Chem.* **42** (1938) , 1001.
[17] D. O. Nierderhauser and F. E. Bartell, Report of Progress, Fundamental Research on Occurence of Petroleum, Publication of the American Petroleum Institute, Lord Baltimore Press, Baltimore, Md., p. 114, 1950.
[18] P. L. du Noüy, *J. Gen. Physiol.* **1** (1919), 521.
[19] W. D. Harkins, and H. F. Jordan, *J. Amer. Chem. Soc.* **52** (1930), 1715.
[20] B. B. Freud and H. Z. Freud, *J. Amer. Chem. Soc.* **52** (1930), 1772.
[21] W. D. Harkins and F. E. Brown, *J. Amer. Chem. Soc.* **41** (1919), 499.
[22] J. L. Lando, and H. T. Oakley, *J. Colloid Interface Sci.* **25** (1967), 526.
[23] M. C. Wilkinson and R. L. Kidwell, *J. Colloid Interface Sci.* **35** (1971), 114.
[24] B. Vonnegut, B., *New Sci. Intrum.* **13** (1942), 6.

5 Mechanism of emulsification and the role of the emulsifier

5.1 Introduction

As mentioned in Chapter 1, to prepare an emulsions oil, water, surfactant and energy are needed. The composition of the system and its nature (oil-in-water, O/W, or water-in-oil, W/O) is determined by the nature of the emulsifier and the process applied [1–5]. Parameters such as the volume fraction of the disperse phase, φ, and the droplet size distribution are determined by the composition of the emulsifier layer around the droplets as well as the process of emulsification (see Chapter 6). In addition, the emulsifier composition and its nature determine the physical stability of the emulsion such as its flocculation behaviour, Ostwald ripening and coalescence. As discussed in Chapter 2, emulsion formation is non-spontaneous, and the system is thermodynamically unstable. The kinetic stability of the emulsion is determined by the balance of attractive and repulsive forces (see Chapter 3). It is important to know the process of emulsion formation and the mechanism of emulsification. The role of the emulsifier in droplet deformation and break-up must be considered at a fundamental level. All these factors are discussed below.

5.2 Mechanism of emulsification

This can be considered from a consideration of the energy required to expand the interface, $\Delta A \gamma$ (where ΔA is the increase in interfacial area when the bulk oil with area A_1 produces a large number of droplets with area A_2; $A_2 \gg A_1$ and γ is the interfacial tension). Since γ is positive, the energy to expand the interface is large and positive; this energy term cannot be compensated by the small entropy of dispersion $T\Delta S^{conf}$ (which is also positive) and the total free energy of formation of an emulsion, ΔG^{form} given by equation (5.1) is positive:

$$\Delta G^{form} = \Delta A \gamma_{12} - T\Delta S^{conf}. \tag{5.1}$$

Thus, emulsion formation is non-spontaneous, and energy is required to produce the droplets.

The formation of large droplets (few μm) as is the case for macro-emulsions is fairly easy and hence high speed stirrers such as the Ultraturrax or Silverson Mixer is sufficient to produce the emulsion. In contrast the formation of small drops (submicron as is the case with nano-emulsions) is difficult and this requires a large amount of surfactant and/or energy. The high energy required for formation of nano-emulsions can be understood from a consideration of the Laplace pressure Δp (the difference in pressure between inside and outside the droplet) as given by equations

(5.2) and (5.3):

$$\Delta p = \gamma \left(\frac{1}{r_1} + \frac{1}{r_2} \right) ,$$ (5.2)

where r_1 and r_2 are the two principal radii of curvature.

For a perfectly spherical droplet $r_1 = r_2 = r$ and

$$\Delta p = \frac{2\gamma}{r} .$$ (5.3)

To break up a drop into smaller ones, it must be strongly deformed, and this deformation increases Δp. This is illustrated in Fig. 5.1, which shows the situation when a spherical drop deforms into a prolate ellipsoid [2].

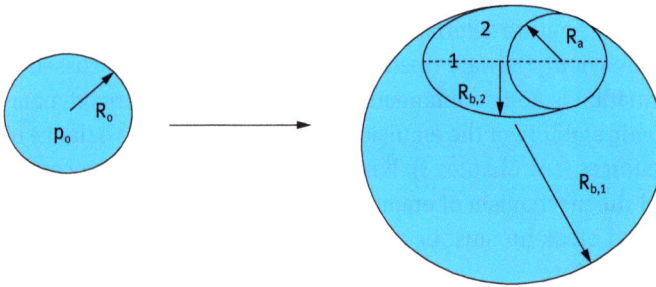

Fig. 5.1: Illustration of increase in Laplace pressure when a spherical drop is deformed to a prolate ellipsoid.

Near 1 there is only one radius of curvature R_o, whereas near 2 there are two radii of curvature $R_{b,1}$ and $R_{b,2}$. The Laplace pressure near 1 is thus larger than that near 2 and also larger than in the undeformed drop. Consequently, the stress needed to deform the drop is higher for a smaller drop. Since the stress is generally transmitted by the surrounding liquid via agitation, higher stresses need more vigorous agitation, and hence more energy is needed to produce smaller drops.

Surfactants play major roles in the formation of emulsions. By lowering the interfacial tension, Δp is reduced and hence the stress needed to break up a drop is reduced. Surfactants also prevent coalescence of newly formed drops (see below).

Fig. 5.2 shows an illustration of the various processes occurring during emulsification: break up of droplets, adsorption of surfactants and droplet collision (which may or may not lead to coalescence) [2].

Each of the above processes occurs numerous times during emulsification, and the time scale of each process is very short, typically a microsecond. This shows that the emulsification process is a dynamic process, and events that occur in a microsecond range could be very important.

To describe emulsion formation one has to consider two main factors: hydrodynamics and interfacial science. In hydrodynamics one has to consider the type of flow:

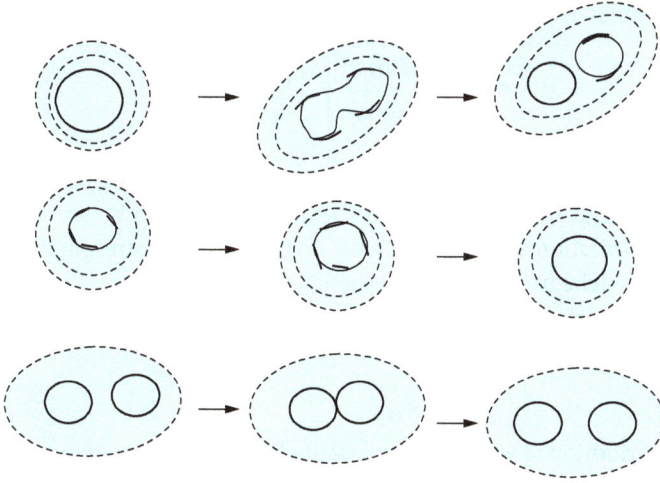

Fig. 5.2: Schematic representation of the various processes occurring during emulsion formation. The drops are depicted by thin lines and the surfactant by heavy lines and dots.

laminar flow and turbulent flow. This depends on the Reynolds number, as will be discussed in Chapter 6.

To assess emulsion formation, one usually measures the droplet size distribution, using, for example, laser diffraction techniques. If the number frequency of droplets as a function of droplet diameter d is given by $f(d)$, the n-th moment of the distribution is

$$S_n = \int_0^\infty d^n f(d) \partial d .$$ (5.4)

The mean droplet size is defined as the ratio of selected moments of the size distribution:

$$d_{nm} = \left[\frac{\int_0^\infty d^n f(d) \partial d}{\int_0^\infty d^m f(d) \partial d} \right]^{1/(n-m)} ,$$ (5.5)

where n and m are integers, and $n > m$ and typically n does not exceed 4.

Using equation (5.5) one can define several mean average diameters:
–   the Sauter mean diameter with $n = 3$ and $m = 2$:

$$d_{32} = \left[\frac{\int_0^\infty d^3 f(d) \partial d}{\int_0^\infty d^2 f(d) \partial d} \right] ;$$ (5.6)

–   the mass mean diameter:

$$d_{43} = \left[\frac{\int_0^\infty d^4 f(d) \partial d}{\int_0^\infty d^3 f(d) \partial d} \right] ;$$ (5.7)

– the number mean diameter:

$$d_{10} = \left[\frac{\int_0^\infty d^1 f(d) \partial d}{\int_0^\infty f(d) \partial d} \right].$$
(5.8)

In most cases d_{32} (the volume/surface average or Sauter mean) is used. The width of the size distribution can be given as the variation coefficient c_m, which is the standard deviation of the distribution weighted with d_m divided by the corresponding average d. Generally a C_2 will be used which corresponds to d_{32}.

Another is the specific surface area A (surface area of all emulsion droplets per unit volume of emulsion):

$$A = \pi S_2 = \frac{6\varphi}{d_{32}}.$$
(5.9)

A typical droplet size distribution of an emulsion measured using the light diffraction technique (Malvern Master sizer) is shown in Fig. 5.3.

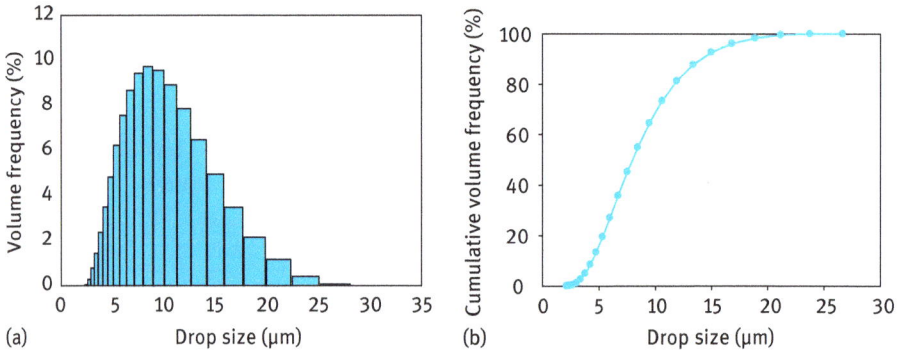

Fig. 5.3: Droplet size distribution of an emulsion: (a) volume frequency in discrete classes; (b) cumulative volume distribution.

5.3 Role of surfactants in emulsion formation

5.3.1 Role of surfactants in reduction of droplet size

Surfactants lower the interfacial tension γ, and this causes a reduction in droplet size. The latter decreases with the decrease in γ. For laminar flow the droplet diameter is proportional to γ; for turbulent inertial regime, the droplet diameter is proportional to $\gamma^{3/5}$.

The surfactant can lower the interfacial tension γ_o of a clean oil-water interface to a value γ, and

$$\pi = \gamma_o - \gamma,$$
(5.10)

where π is the surface pressure. The dependence of π on the surfactant activity a or concentration C is given by the Gibbs equation, as discussed in Chapter 4:

$$d\pi = -d\gamma = RT\,\Gamma\,d\ln a = RT\,\Gamma\,d\ln C, \tag{5.11}$$

where R is the gas constant, T is the absolute temperature and Γ is the surface excess (number of moles adsorbed per unit area of the interface).

At high a, the surface excess Γ reaches a plateau value; for many surfactants it is of the order of 3 mg m^{-2}. Γ increases with an increase in surfactant concentration, and eventually it reaches a plateau value (saturation adsorption). This is illustrated in Fig. 5.4 for various emulsifiers.

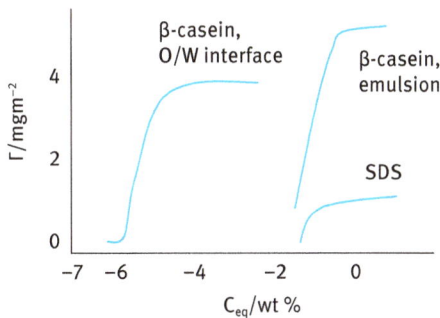

Fig. 5.4: Variation of Γ (mg m^{-2}) with log C_{eq} / wt %. The oils are β-casein (O/W interface) toluene, β-casein (emulsions) soybean, SDS benzene.

It can be seen from Fig. 5.4 that the polymer (β-casein) is more surface active than the surfactant (SDS). The value of C needed to obtain the same Γ is much smaller for the polymer when compared with the surfactant. In contrast, the value of γ reached at full saturation of the interface is lower for a surfactant (mostly in the region of 1–3 mN m^{-1}, depending on the nature of surfactant and oil) when compared with a polymer (with γ values in the region of 10–20 mN m^{-1}, depending on the nature of polymer and oil). This is due to the much closer packing of the small surfactant molecules at the interface when compared with the much larger polymer molecule that adopts tail-train-loop-tail conformation.

The effect of reducing γ on the droplet size is illustrated in Fig. 5.5, which shows a plot of the droplet surface area A and mean drop size d_{32} as a function of surfactant concentration m for various systems. The amount of surfactant required to produce the smallest drop size will depend on its activity a (concentration) in the bulk, which determines the reduction in γ, as discussed above.

Another important role of the surfactant is its effect on the interfacial dilational modulus ε:

$$\varepsilon = \frac{d\gamma}{d\ln A}. \tag{5.12}$$

ε is the absolute value of a complex quantity, composed of an elastic and viscous term.

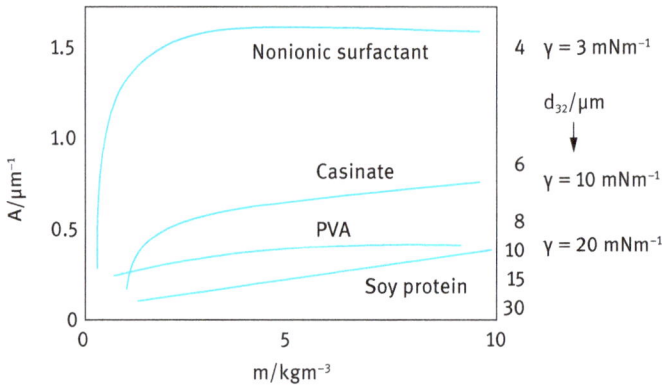

Fig. 5.5: Variation of A and d_{32} with m for various surfactant systems.

During emulsification an increase in the interfacial area A takes place, and this causes a reduction in Γ. The equilibrium is restored by adsorption of surfactant from the bulk, but this takes time (shorter times occur at higher surfactant activity). Thus ε is small at small a and also at large a. Because of the lack or slowness of equilibrium with polymeric surfactants, ε will not be the same for the expansion and compression of the interface.

In practice, emulsifiers are generally made of surfactant mixtures, often containing different components; these have pronounced effects on γ and ε. Some specific surfactant mixtures give lower γ values than either of the two individual components. The presence of more than one surfactant molecule at the interface tends to increase ε at high surfactant concentrations. The various components vary in surface activity. Those with the lowest γ tend to predominate at the interface, but if present at low concentrations, it may take long time before reaching the lowest value. Polymer-surfactant mixtures may show some synergetic surface activity.

During emulsification, surfactant molecules are transferred from the solution to the interface, and this leaves an ever lower surfactant activity [2]. Consider for example an O/W emulsion with a volume fraction $\varphi = 0.4$ and a Sauter diameter $d_{32} = 1\,\mu m$. According to equation (5.9), the specific surface area is $2.4\,m^2\,ml^{-1}$ and for a surface excess Γ of $3\,mg\,m^{-2}$, the amount of surfactant at the interface is $7.2\,mg\,ml^{-1}$ emulsion, corresponding to $12\,mg\,ml^{-1}$ aqueous phase (or 1.2%). Assuming that the concentration of surfactant, C_{eq} (the concentration left after emulsification), leading to a plateau value of Γ equals $0.3\,mg\,ml^{-1}$, then the surfactant concentration decreases from 12.3 to $0.3\,mg\,ml^{-1}$ during emulsification. This implies that the effective γ value increases during the process. If insufficient surfactant is present to leave a concentration C_{eq} after emulsification, even the equilibrium γ value would increase.

Another aspect is that the composition of surfactant mixture in solution may alter during emulsification [2]. If some minor components are present which give a relatively small γ value, this will predominate at a macroscopic interface, but during emulsification, as the interfacial area increases the solution will soon become depleted of these components. Consequently, the equilibrium value of γ will increase during the process, and the final value may be markedly larger than what is expected on the basis of the macroscopic measurement.

5.3.2 Role of surfactants in droplet deformation

During droplet deformation, its interfacial area is increased [2]. The drop will commonly have acquired some surfactant, and it may even have a Γ value close to the equilibrium at the prevailing (local) surface activity. The surfactant molecules may distribute themselves evenly over the enlarged interface by surface diffusion or by spreading. The rate of surface diffusion is determined by the surface diffusion coefficient D_s, inversely proportional to the molar mass of the surfactant molecule, and also inversely proportional to the effective viscosity felt. D_s also decreases with increase of Γ. Sudden extension of the interface or sudden application of a surfactant to an interface can produce a large interfacial tension gradient, and in such a case spreading of the surfactant can occur.

Surfactants allow the existence of interfacial tension gradients which is crucial for the formation of stable droplets. In the absence of surfactants (clean interface), the interface cannot withstand a tangential stress; the liquid motion will be continuous across a liquid interface (Fig. 5.6 (a)).

If a liquid flows along the interface with surfactants, the latter will be swept downstream, causing an interfacial tension gradient (Fig. 5.6 (b)). A balance of forces will be established:

$$\eta \left[\frac{dV_x}{dy} \right]_{y=0} = -\frac{dy}{dx} . \tag{5.13}$$

If the γ-gradient can become large enough, it will arrest the interface. The largest value attainable for $d\gamma$ equals about π_{eq}, i.e. $\gamma_o - \gamma_{eq}$. If it acts over a small distance, a considerable stress can develop, of the order of 10 kPa.

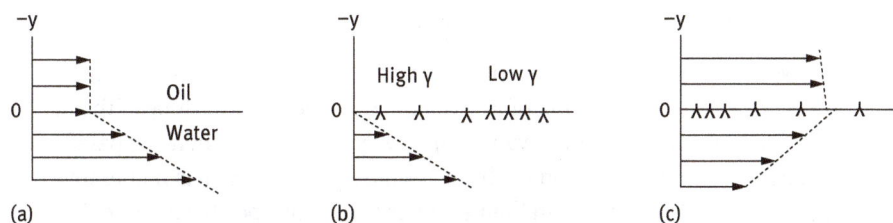

Fig. 5.6: Interfacial tension gradients and flow near an oil/water interface: (a) no surfactant; (b) velocity gradient causes an interfacial tension gradient; (c) interfacial tension gradient causes flow (Marangoni effect).

If the surfactant is applied at one site of the interface, a γ-gradient is formed which will cause the interface to move roughly at a velocity given by

$$v = 1.2[\eta\rho z]^{-1/3}|\Delta\gamma|^{2/3} . \tag{5.14}$$

The interface will then drag some of the bordering liquid with it (Fig. 5.6 (c)). This is called the Marangoni effect [2].

Interfacial tension gradients are very important in stabilising the thin liquid film between the droplets, which is very important during the beginning of emulsification, when films of the continuous phase may be drawn through the disperse phase, or when collision of the still large deformable drops causes the film to to form between them. The magnitude of the γ-gradients and of the Marangoni effect depends on the surface dilational modulus ε, which for a plane interface with one surfactant-containing phase, is given by the expressions

$$\varepsilon = \frac{-d\gamma/d\ln\Gamma}{(1 + 2\xi + 2\xi^2)^{1/2}} , \tag{5.15}$$

$$\xi = \frac{dm_C}{d\Gamma}\left(\frac{D}{2\omega}\right)^{1/2} , \tag{5.16}$$

$$\omega = \frac{d\ln A}{dt} , \tag{5.17}$$

where D is the diffusion coefficient of the surfactant, and ω represents a time scale (time needed for doubling the surface area) roughly equal to τ_{def}.

During emulsification, ε is dominated by the magnitude of the numerator in equation (5.15) because ξ remains small. The value of $dm_C/d\Gamma$ tends to go to very high values when Γ reaches its plateau value; ε goes to a maximum when m_C is increased. However, during droplet deformation, Γ will always remain smaller. Taking reasonable values for the variables; $dm_C/d\Gamma = 10^2{-}10^4\,m^{-1}$, $D = 10^{-9}{-}10^{-11}\,m^2\,s^{-1}$ and $\tau_{def} = 10^{-2}{-}10^{-6}$ s, ξ < 0.1 at all conditions. The same conclusion can be drawn for values of ε in thin films, e.g. between closely approaching drops. It may be concluded that for conditions that prevail during emulsification, ε increases with m_C and follows the relation

$$\varepsilon \approx \frac{d\pi}{d\ln\Gamma} \tag{5.18}$$

except for very high surfactant concentration, where π is the surface pressure (π = $\gamma_o - \gamma$). Fig. 5.7 shows the variation of π with ln Γ; ε is given by the slope of the line.

The SDS shows a much higher ε value during emulsification, when compared with the polymers β-casein and lysozome. This is because the value of Γ is higher for SDS. The two proteins show difference in their ε values which may be attributed to the conformational change that occur upon adsorption.

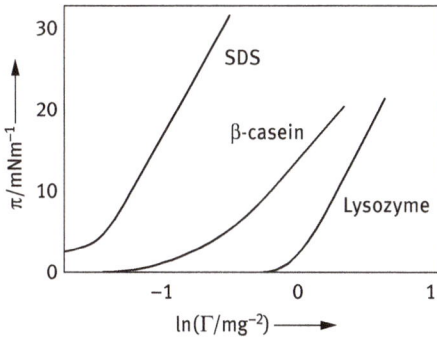

Fig. 5.7: π versus ln Γ for various emulsifiers.

The presence of a surfactant means that during emulsification the interfacial tension need not be the same everywhere (see Fig. 5.6). This has two consequences: (i) the equilibrium shape of the drop is affected; (ii) any γ-gradient formed will slow down the motion of the liquid inside the drop (this diminishes the amount of energy needed to deform and break up the drop).

Another important role of the emulsifier is to prevent coalescence during emulsification. This is certainly not due to the strong repulsion between the droplets, since the pressure at which two drops are pressed together is much greater than the repulsive stresses. The counteracting stress must be due to the formation of γ-gradients. When two drops are pushed together, liquid will flow out from the thin layer between them, and the flow will induce a γ-gradient. This was shown in Fig. 5.6 (c) This produces a counteracting stress given by

$$\tau_{\Delta\gamma} \approx \frac{2|\Delta\gamma|}{(1/2)d}. \tag{5.19}$$

The factor 2 follows from the fact that two interfaces are involved. Taking a value of $\Delta\gamma = 10\,\mathrm{mN\,m^{-1}}$, the stress amounts to 40 kPa (which is of the same order of magnitude as the external stress). The stress due to the γ-gradient cannot as such prevent coalescence, since it only acts for a short time, but it will greatly slow down the mutual approach of the droplets. The external stress will also act for a short time, and it may well be that the drops move apart before coalescence can occur. The effective γ-gradient will depend on the value of ε as given by equation (5.18).

Closely related to the above mechanism, is the Gibbs–Marangoni effect [6–8], schematically represented in Fig. 5.8. The depletion of surfactant in the thin film between approaching drops results in γ-gradient without liquid flow being involved. This results in an inward flow of liquid that tends to drive the drops apart. Such a mechanism would only act if the drops are insufficiently covered with surfactant (Γ below the plateau value), as occurs during emulsification.

The Gibbs–Marangoni effect also explains the Bancroft rule, which states that the phase in which the surfactant is most soluble form the continuous phase. If the surfactant is in the droplets, a γ-gradient cannot develop, and the drops would be

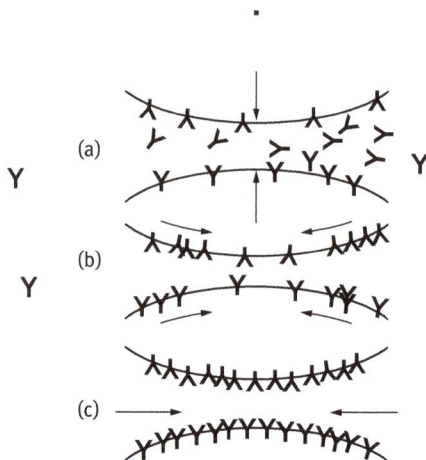

Fig. 5.8: Schematic representation of the Gibbs–Marangoni effect for two approaching drops.

prone to coalescence. Thus, surfactants with HLB > 7 tend to form O/W emulsions, and HLB < 7 tend to form W/O emulsions.

The Gibbs–Marangoni effect also explains the difference between surfactants and polymers for emulsification. Polymers give larger drops when compared with surfactants. Polymers give a smaller value of ε at small concentrations when compared to surfactants (Fig. 5.7).

Various other factors should also be considered for emulsification, as will be discussed in detail in Chapter 6: the disperse phase volume fraction φ. An increase in φ leads to an increase in droplet collision and hence coalescence during emulsification. With an increase in φ, the viscosity of the emulsion increases and could change the flow from turbulent to laminar. The presence of many particles results in a local increase in velocity gradients. In turbulent flow, increase in φ will induce turbulence depression (see Chapter 6). This will result in larger droplets. Turbulence depression by added polymers tend to remove the small eddies, resulting in the formation of larger droplets.

If the mass ratio of surfactant to continuous phase is kept constant, an increase in φ results in a decrease in surfactant concentration, and hence an increase in γ_{eq} results in larger droplets. If the mass ratio of surfactant to disperse phase is kept constant, the above changes are reversed.

References

[1] Th. F. Tadros and B. Vincent, in: P. Becher (ed), *Encyclopedia of Emulsion Technology*, Marcel Dekker, New York, 1983.
[2] P. Walstra and P. E. A. Smolders, in: B. P. Binks (ed.), *Modern Aspects of Emulsions*, The Royal Society of Chemistry, Cambridge, 1998.
[3] Tharwat Tadros, *Applied Surfactants*, Wiley-VCH, Germany, 2005.

[4] Tharwat Tadros, *Emulsion Formation Stability and Rheology*, in: Tharwat Tadros (ed.), *Emulsion Formation and Stability*, Ch. 1, Wiley-VCH, Germany, 2013.
[5] T. Tadros (ed.), *Encyclopedia of Colloid and Interface Science*, Springer, Berlin, 2013.
[6] E. H. Lucasses-Reynders, *Colloids and Surfaces* **A91** (1994), 79.
[7] J. Lucassen, in: E. H. Lucassesn-Reynders (ed.), *Anionic Surfactants*, Marcel Dekker, New York, 1981.
[8] M. van den Tempel, Proc. Int. Congr. Surf. Act. **2** (1960) 573.

6 Methods of emulsification

6.1 Introduction

Several procedures may be applied for emulsion preparation, these range from simple pipe flow (low agitation energy L), static mixers (toothed devices such as the Ultra-Turrax and batch radial discharge mixers such as the Silverson mixers) and general stirrers (low to medium energy, L–M), colloid mills and high pressure homogenizers (high energy, H), ultrasound generators (M–H) and membrane emulsification methods. The method of preparation can be continuous (C) or batch-wise (B): Pipe flow – C; static mixers and general stirrers – B, C; colloid mill and high pressure homogenizers – C; ultrasound – B, C.

In all these methods, there is liquid flow: unbounded and strongly confined flow. In the unbounded flow any droplets is surrounded by a large amount of flowing liquid (the confining walls of the apparatus are far away from most of the droplets). The forces can be frictional (mostly viscous) or inertial. Viscous forces cause shear stresses to act on the interface between the droplets and the continuous phase (primarily in the direction of the interface). The shear stresses can be generated by laminar flow (LV) or turbulent flow (TV); this depends on the dimensionless Reynolds numbers Re:

$$Re = \frac{v l \rho}{\eta},$$
(6.1)

where v is the linear liquid velocity, ρ is the liquid density and η is its viscosity. l is a characteristic length that is given by the diameter of flow through a cylindrical tube and by twice the slit width in a narrow slit.

For laminar flow $Re < \sim 1000$, whereas for turbulent flow $Re > \sim 2000$. Thus, whether the regime is linear or turbulent depends on the scale of the apparatus, the flow rate and the liquid viscosity [1–4].

Below, a brief description of the most commonly used methods of emulsification is given, followed by analysis of the Laminar and turbulent flow.

6.2 Rotor-stator mixers

These are the most commonly used mixers for emulsification. Two main types are available and these are described below [5].

6.2.1 Toothed devices

The most commonly used toothed device (schematically illustrated in Fig. 6.1) is the Ultra-Turrax (IKA works, Germany).

Fig. 6.1: Schematic representation
of a toothed mixer (Ultra-Turrax).

Toothed devices are available both as in-line as well as batch mixers, and because of their open structure they have a relatively good pumping capacity: therefore in batch applications they frequently do not need an additional impeller to induce bulk flow, even in relatively large mixing vessels. These mixers are used in the food industry to manufacture ice cream, margarine and salad dressings, in cosmetic and personal care products to manufacture creams and lotions, as well as in the manufacture of speciality chemicals for micro-encapsulation of waxes and paraffin. They are also popular in the paper industry to process highly viscous and non-Newtonian paper pulp, and in the manufacturing of paints and coatings. Ultra-Turrax mixers have been used to manufacture emulsion-based lipid carriers with drops below 1 µm and in emulsion polymerisation to produce drops of the order of 300 nm.

6.2.2 Batch radial discharge mixers

Batch radial discharge mixers such as Silverson mixers (Fig. 6.2) have a relatively simple design with a rotor equipped with four blades pumping the fluid through a stationary stator perforated with differently shaped/sized holes or slots.

They are frequently supplied with a set of easily interchangeable stators enabling the same machine to be used for a range of operations such as emulsification, homogenization, blending, particle size reduction and deagglomeration. Changing from one screen to another is quick and simple. Different stators/screens used in batch Silverson mixers are shown in Fig. 6.3. The general purpose disintegrating stator (Fig. 6.3 (a)) is recommended for preparation of thick emulsions (gels) whilst the slotted disintegrating stator (Fig. 6.3 (b)) is designed for emulsions containing elastic materials such as polymers. Square hole screens (Fig. 6.3 (c)) are recommended for the preparation of

Fig. 6.2: Schematic representation of batch radial discharge mixer (Silverson mixer).

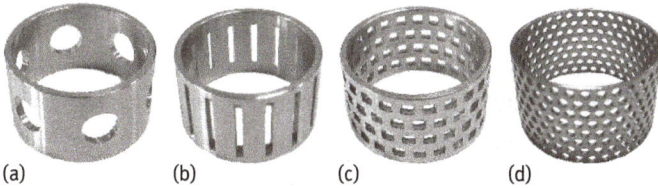

(a)   (b)   (c)   (d)

Fig. 6.3: Stators used in batch Silverson radial discharge mixers.

emulsions whereas the standard emulsor screen (Fig. 6.3 (d)) is used for liquid/liquid emulsification.

Radial discharge high shear mixers are used in a wide range of industries ranging from foods through to chemicals, cosmetics and pharmaceutical industries. Silverson rotor-stator mixers are used in cosmetic and pharmaceutical industries to manufacture both concentrated liquid-liquid and liquid-solid emulsions such as creams, lotions, mascaras and deodorants to name the most common applications.

6.2.3 Design and arrangement

Batch toothed and radial discharge rotor-stator mixers are manufactured in different sizes ranging from the laboratory to the industrial scale. In lab applications mixing heads (assembly of rotor and stator) can be as small as 0.01 m (Turrax, Silverson) and the volume of processed fluid can vary from several millilitres to few litres. In models used in industrial applications mixing heads might have up to 0.5 m diameter enabling processing of several cubic meters of fluids in one batch.

In practical applications the selection of the rotor-stator mixer for a specific emulsification process depends on the required morphology of the product, frequently quantified in terms of average drop size or in terms of drop size distributions, and by the scale of the process. There is very little information enabling calculation of average drop size in rotor-stator mixers and there are no methods enabling estimation of drop size distributions. Therefore the selection of an appropriate mixer and processing conditions for a required formulation is frequently carried out by trial and error. Initially, one can carry out lab scale emulsification of given formulations testing different type/geometries of mixers they manufacture. Once the type of mixer and its operating parameters are determined at the lab scale the process needs to be scaled up. The majority of lab tests of emulsification is carried out in small batch vessels, as it is easier and cheaper than running continuous processes therefore prior to scaling up of the rotor-stator mixer it has to be decided whether industrial emulsification should be run as a batch or as a continuous process. Batch mixers are recommended for processes where formulation of a product requires long processing times typically associated with slow chemical reactions. They require simple control systems, but spatial homogeneity may be an issue in large vessels which could lead to a longer processing time. In processes where quality of the product is controlled by mechanical/hydrodynamic interactions between continuous and dispersed phases or by fast chemical reactions, but large amounts of energy is necessary to ensure adequate mixing, in-line rotor-stator mixers are recommended. In-line mixers are also recommended to efficiently process large volumes of fluid.

In the case of batch processing, rotor-stator devices immersed as top entry mixers is mechanically the simplest arrangement, but in some processes bottom entry mixers ensures better bulk mixing; however in this case sealing is more complex. In general, the efficiency of batch rotor-stator mixers decreases as the vessel size increases and as the viscosity of the processed fluid increases because of limited bulk mixing by rotor-stator mixers. While the open structure of Ultra-Turrax mixers frequently enables sufficient bulk mixing even in relatively large vessels, if the liquid/emulsion has a low apparent viscosity, processing of very viscous emulsions requires an additional impeller (typically anchor type) to induce bulk flow and to circulate the emulsion through the rotor-stator mixer. On the other hand, batch Silverson rotor-stator mixers have a very limited pumping capacity, and even at the lab scale they are mounted off the centre of the vessel to improve bulk mixing. At the large scale there is always need for at least one additional impeller, and in the case of very large units more than one impeller is mounted on the same shaft.

Problems associated with the application of batch rotor-stator mixers for processing large volumes of fluid discussed above can be avoided by replacing batch mixers with in-line (continuous) mixers. There are many designs offered by different suppliers (Silverson, IKA, etc.), and the main differences are related to the geometry of the rotors and stators with stators and rotors designed for different applications. The main difference between batch and in-line rotor-stator mixers is that the latter have a strong pumping capacity and are therefore mounted directly in the pipeline. One of the main advantages of in-line over batch mixers is that for the same power demand a much smaller mixer is required; they are therefore better suited for processing large volumes of fluid. When the scale of the processing vessel increases, a point is reached where it is more efficient to use an in-line rotor-stator mixer rather than a batch mixer with a large diameter. Because power consumption increases sharply with rotor diameter (to the fifth power), an excessively large motor is necessary at large scales. This transition point depends on the fluid rheology, but for a fluid with a viscosity similar to water it is recommended to change from a batch to an in-line rotor-stator process at a volume of approximately 1 to 1.5 tonnes. The majority of manufacturers supply both single and multistage mixers with the emulsification of highly viscous liquids.

6.3 Flow regimes

As mentioned above, in all methods there is liquid flow, unbounded flow and strongly confined flow. In the unbounded flow any droplets are surrounded by a large amount of flowing liquid (the confining walls of the apparatus are far away from most of the droplets); the forces can be frictional (mostly viscous) or inertial. Viscous forces cause shear stresses to act on the interface between the droplets and the continuous phase (primarily in the direction of the interface). The shear stresses can be generated by laminar flow (LV) or turbulent flow (TV); This depends on the Reynolds number Re, as given by equation (6.1). For laminar flow Re $< \sim 1000$, whereas for turbulent flow Re $> \sim 2000$. Thus whether the regime is linear or turbulent depends on the scale of the apparatus, the flow rate and the liquid viscosity. If the turbulent eddies are much larger than the droplets, they exert shear stresses on the droplets. If the turbulent eddies are much smaller than the droplets, inertial forces will cause disruption (TI). In bounded flow other relations hold; if the smallest dimension of the part of the apparatus in which the droplets are disrupted (say a slit) is comparable to droplet size, other relations hold (the flow is always laminar. A different regime prevails if the droplets are directly injected through a narrow capillary into the continuous phase (Injection regime), i.e. membrane emulsification.

Within each regime, an essential variable is the intensity of the forces acting; the viscous stress during laminar flow $\sigma_{viscous}$ is given by

$$\sigma_{viscous} = \eta G, \qquad (6.2)$$

where G is the velocity gradient.

The intensity in turbulent flow is expressed by the power density ε (the amount of energy dissipated per unit volume per unit time); for turbulent flow,

$$\varepsilon = \eta G^2. \qquad (6.3)$$

The most important regimes are: laminar/viscous (LV); turbulent/viscous (TV); turbulent/inertial (TI). For water as the continuous phase, the regime is always TI. For higher viscosity of the continuous phase ($\eta_C = 0.1\,Pa\,s$), the regime is TV. For still higher viscosity or a small apparatus (small l), the regime is LV. For very small apparatus (as is the case with most laboratory homogenizers), the regime is nearly always LV.

Tab. 6.1: Various regimes for emulsification.

Regime	Laminar	Turbulent	Turbulent
	Viscous LV	Viscous TV	Inertial TI
Re_{flow}	$< \sim 1000$	$> \sim 2000$	$> \sim 2000$
Re_{drop}	< 1	< 1	> 1
$\sigma_{ext} \sim$	$\eta_C G$	$\varepsilon^{1/2} \eta_C^{1/2}$	$\varepsilon^{2/3} d^{2/3} \rho^{1/3}$
$d \sim$	$(2\gamma We_{cr} \eta_C G)$	$(\gamma / \varepsilon^{1/2} \eta_C^{1/2})$	$(\gamma^{3/5} / \varepsilon^{2/5} \rho^{1/5})$
τ_{def}	$(\eta_D / \eta_C G)$	$(\eta_D / \varepsilon^{1/2} \eta_C^{1/2})$	$(\eta_D / \varepsilon^{2/3} d^{2/3} \rho^{1/3})$
$\tau_{ads} \sim$	$(6\pi\Gamma / dm_c G)$	$(6\pi\Gamma\eta_C / dm_c \varepsilon^{1/2})$	$(\Gamma\rho^{1/3} / d^{1/3} m_c \varepsilon^{1/3})$

For the above regimes, a semi-quantitative theory is available that can give the time scale and magnitude of the local stress σ_{ext}, the droplet diameter d, time scale of droplets deformation τ_{def}, time scale of surfactant adsorption, τ_{ads} and mutual collision of droplets, as illustrated in Tab. 6.1.

6.3.1 Laminar flow

Laminar flow can be of a variety of types, purely rotational to purely extensional as illustrated in Fig. 6.4. For simple shear the flow consists of equal parts of rotation and elongation. The velocity gradient G (in reciprocal seconds) is equal to the shear rate γ. For hyperbolic flow G is equal to the elongation rate. The strength of a flow is generally expressed by the stress it exerts on any plane in the direction of flow. It is simply equal to Gη (η is simply the shear viscosity).

For elongational flow, the elongational viscosity η_{el} is given by

$$\eta_{el} = Tr \, \eta \, , \tag{6.4}$$

where Tr is the dimensionless Trouton number which is equal to 2 for Newtonian liquids in two-dimensional uneasily elongation flow. Tr = 3 for axisymmetric uniaxial flow and is equal to 4 for biaxial flows. Elongational flows exert higher stresses for the same value of G than for simple shear. For non-Newtonian liquids, the relationships are more complicated, and the values of Tr tends to be much higher.

An important parameter that describes droplet deformation is the Weber number We (which gives the ratio of the external stress over the Laplace pressure):

$$We = \frac{G\eta_C R}{2\gamma} \, . \tag{6.5}$$

The deformation of the drop increases with increase of We and above a critical value We_{cr} the drop bursts form smaller droplets. We_{cr} depends on two parameters: (i) the

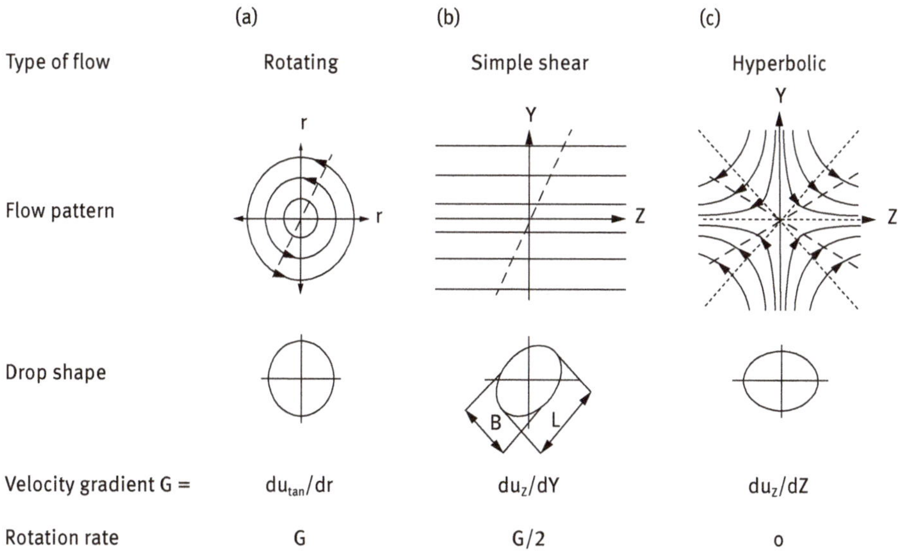

Type of flow	(a) Rotating	(b) Simple shear	(c) Hyperbolic
Flow pattern			
Drop shape			
Velocity gradient G =	du_{tan}/dr	du_z/dY	du_z/dZ
Rotation rate	G	G/2	0

Fig. 6.4: Various types of two-dimensional flow and the effect on droplet Deformation and rotation.

velocity vector α (α = 0 for simple shear and (ii) α = 1 for hyperbolic flow. The viscosity ratio λ of is that of the oil η_D and the external continuous phase η_C:

$$\lambda = \frac{\eta_D}{\eta_C}.\tag{6.6}$$

The variation of critical Weber number with λ at various α values is shown in Fig. 6.5.

As mentioned above, the viscosity of the oil plays an important role in the break-up of droplets; the higher the viscosity, the longer it will take to deform a drop. The deformation time τ_{def} is given by the ratio of oil viscosity to the external stress acting on the drop:

$$\tau_{def} = \frac{\eta_D}{\sigma_{ext}}.\tag{6.7}$$

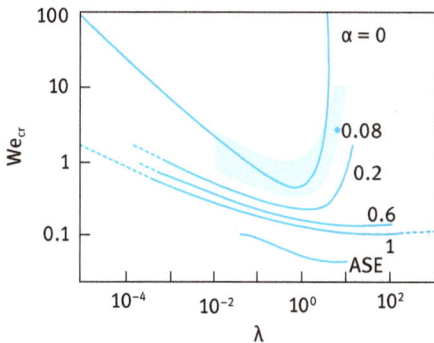

Fig. 6.5: Critical Weber number for the break-up of drops in various types of flow. The hatched area represent the apparent We_{cr} in a colloid mill.

The above ideas for simple laminar flow were tested using emulsions containing 80 % oil in water stabilized with egg yolk. A colloid mill and static mixers were used to prepare the emulsion. The results are shown in Fig. 6.6, which gives the number of droplets n in which a parent drop is broken down when it is suddenly extended into a long thread, corresponding to We_b which is larger than We_{cr}. The number of drops increases with increase of We_b/We_{cr}. The largest number of drops, i.e. the smaller the droplet size is obtained when $\lambda = 1$, i.e. when the viscosity of the oil phase is closer to that of the continuous phase. In practice, the resulting drop size distribution is of greater importance than the critical drop size for break-up.

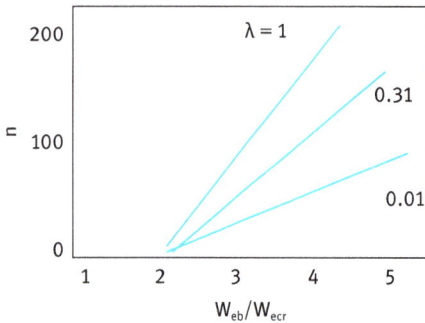

Fig. 6.6: Variation of n with We_b/We_{cr}.

6.3.2 Turbulent flow

Turbulent flow is characterised by the presence of eddies, which means that the average local flow velocity u generally differs from the time average value $\bar{u}$. The velocity fluctuates in a chaotic way, and the average difference between u and u′ equals zero; however, the root mean square average u′ is finite [5–8]:

$$u' = \langle (u - \bar{u})^2 \rangle^{1/2} . \tag{6.8}$$

The value of u′ generally depends on direction, but for very high Re (> 50 000) and at small length scales the turbulence flow can be isotropic, and u′ does not depend on direction. Turbulent flow shows a spectrum of eddy sizes (l); the largest eddies have the highest u′, and they transfer their kinetic energy to smaller eddies, which have a smaller u′ but larger velocity gradient u′/l.

The smallest eddy size is given by

$$l_e = \eta^{3/4} \rho^{-1/2} \varepsilon^{-1/4} . \tag{6.9}$$

ε is called the power density, i.e. the amount of energy dissipated per unit volume of liquid and per unit time (given in $W\,m^{-3}$).

The local flow velocities depend on the distance scale x considered, and for a scale comparable to the size of an energy-bearing eddy (x ~ l_e). The velocity near that eddy is given by

$$u'(x) = \varepsilon^{1/3} x^{1/3} \rho^{-1/3} . \tag{6.10}$$

The velocity gradient in an eddy is given by $u'(x)/(x)$, and it increases strongly with decreasing size. The eddies have a short lifetime, given by

$$\tau(l_e) = l_e/u'(l_e) = l_e^{2/3} \varepsilon^{-1/3} \rho^{1/3} . \tag{6.11}$$

The local flow velocity for scales much smaller than the size of an energy-bearing eddy $(x \ll l_e)$ is given by

$$u'(x) = \varepsilon^{1/2} x \eta^{-1/2} . \tag{6.12}$$

The break-up of droplets in turbulent flow due to inertial forces may be represented by local pressure fluctuations near energy-bearing eddies:

$$\Delta p(x) = \rho[u'(x)]^2 = \varepsilon^{2/3} x^{2/3} \rho^{1/3} , \tag{6.13}$$

where ε is the power density, i.e. the amount of energy dissipated per unit volume, ρ is the density and x is the distance scale. If Δp is larger than the Laplace pressure $(p = 2\gamma/R)$ near the eddy, the drop would be broken up. Break-up would be most effective if $d = l_e$.

Putting $x = d_{max}$, the following expression gives the largest drops that are not broken up in the turbulent field:

$$d_{max} = \varepsilon^{-2/5} \gamma^{3/5} \rho^{-1/5} . \tag{6.14}$$

The validity of equation (6.14) is subject to two conditions: (i) the droplet size obtained cannot be much smaller than l_o – this equation is fulfilled for small η_C; (ii) the flow near the droplet should be turbulent – this depends on the droplet Reynolds number given by

$$Re_{dr} = du'(d)\rho_C/\eta_C . \tag{6.15}$$

The condition $Re_{dr} > 1$ and combination with equation (6.10) leads to

$$d > \eta_C^2/\gamma\rho . \tag{6.16}$$

Provided that (i) φ is small, (ii) η_C is not much larger than $1\,\mathrm{mPa\,s}$, (iii) η_D is fairly small, (iv) γ is constant, and (v) the machine is fairly small, equation (6.14) seems to hold well even for non-isotropic turbulence with the Reynolds number much smaller than 50 000. The smallest drops are produced at the highest power density. Since the power density varies from place to place (particularly if Re is not very high), the droplet size distribution can be very wide. For break-up of drops in TI regime, the flow near the drop is turbulent. For laminar flow, break-up by viscous forces is possible. If the flow rate near the drop (u) varies greatly with distance d, the local velocity gradient is G. A pressure difference is produced over the drop of $\frac{1}{2}\Delta\rho(u_2) = \rho Gd$. At the same time a shear stress $\eta_C G$ acts on the drop. The viscous forces will be predominant for $\eta_C G > \rho G d$, leading to the condition

$$\bar{u} d \rho / \eta_C = Re_{dr} < 1. \tag{6.17}$$

The local velocity gradient is $\eta_C G = \varepsilon^{1/2} \eta_C^{1/2}$. This results in the following expression for d_{max}:

$$d_{max} = We_{cr} \gamma \varepsilon^{-1/2} \eta_C^{-1/2}. \tag{6.18}$$

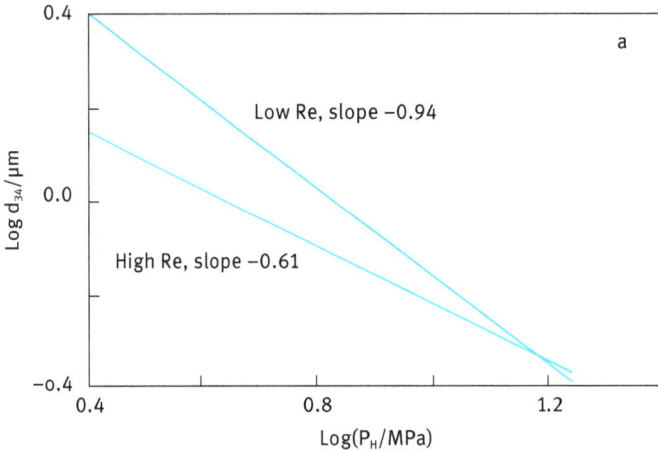

Fig. 6.7: Comparison of droplet size distribution obtained with two high pressure homogenizers; a very small one (low Re) and a large one (high Re).

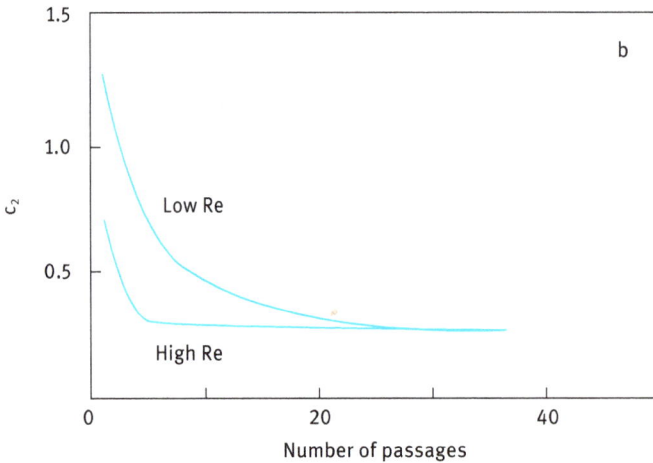

Fig. 6.8: Relative distribution width c_2 as a function of the number of passes through the homogenizers.

The value of We_{cr} is rarely > 1, since the flow has an elongational component. For not very small η_C, d_{max} is smaller for TV than TI.

The viscosity of the oil plays an important role in the break-up of droplets; the higher the viscosity, the longer it will take to deform a drop. The deformation time τ_{def} is given by the ratio of oil viscosity to the external stress acting on the drop:

$$\tau_{def} = \frac{\eta_D}{\eta_C}. \tag{6.19}$$

The viscosity of the continuous phase η_C plays an important role in some regimes: for turbulent inertial regime, η_C has no effect on droplets size. For a turbulent viscous regime, larger η_C leads to smaller droplets. For a laminar viscous one the effect is even stronger.

The value of η_C and the size of the apparatus determines what regime prevails via the effect on Re. In a large machine and low η_C, Re is always very large, and the resulting average droplet diameter d is proportional to $P_H^{-0.6}$ (where P_H is the homogenization pressure). If η_C is higher and $Re_{dr} < 1$, the regime is TV and $d \propto P_H^{-0.75}$. For a smaller machine, as used in the lab, where the slit width of the valve may be of the order of µm, Re is small and the regime is LV; $d \propto P_H^{-1.0}$. If the slit is made very small (of the order of droplet diameter), the regime can become TV.

Fig. 6.7 shows the variation of average droplet diameter d_{43} with P_H at low and high Re for 20 % soybean oil/water emulsion stabilized with sodium caseinate $(30\, mg\, ml^{-1})$.

Fig. 6.8 shows the variation of width of distribution with number of passages at low and high Re.

The addition of high molecular weight polymers in the continuous phase increases η_C, resulting in turbulence depression increasing d_{32} while decreasing c_2.

6.4 Membrane emulsification

In this method, the disperse phase is passed through a membrane, and droplets leaving the pores are immediately taken up by the continuous phase. The membrane is commonly made of porous glass or of ceramic materials. The general configuration is a membrane in the shape of a hollow cylinder; the disperse phase is pressed through it from the outside, and the continuous phase is pumped through the cylinder (cross flow). The flow also causes detachment of the protruding droplets from the membrane [2].

Several requirements are necessary for the process: (i) for a hydrophobic disperse phase (O/W emulsion) the membrane should be hydrophilic, whereas for a hydrophilic disperse phase (W/O emulsion) the membrane should be hydrophobic, since otherwise the droplets cannot be detached; (ii) the pores must be sufficiently far apart to prevent the droplets coming out from touching each other and coalescing; (iii) the pressure over the membrane should be sufficiently high to achieve drop formation. This pressure should be at least of the order of Laplace pressure of a drop of diameter equal to the pore diameter. For example, for pores of 0.4 µm and $\gamma = 5\,\text{mN m}^{-1}$, the pressure should be of the order of 10^5 Pa, but larger pressures are needed in practice; this would amount to $3 \cdot 10^5$ Pa, also to obtain a significant flow rate of the disperse phase through the membrane.

The smallest drop size obtained by membrane emulsification is about three times the pore diameter. The main disadvantage is its slow process, which can be of the order of $10^{-3}\,\text{m}^3$ per m^2 per second. This implies that very long circulation times are needed to produce even small volume fractions.

6.5 Formulation variables and comparison of various emulsification methods

The most important variables that affect the emulsification process are the nature of the oil and emulsifier, the volume fraction of the disperse phase φ and the emulsification process. The effect of the nature of the oil and emulsifier were discussed in detail in Chapter 5. As discussed above, the method of emulsification and the regime (laminar or turbulent) have a pronounced effect on the process and the final droplet size distribution. The effect of the volume fraction of the disperses phase requires special attention. It affects the rate of collision between droplets during emulsification, and thereby the rate of coalescence. As a first approximation, this would depend on the relation between τ_{ads} and τ_{coal} (where τ_{ads} is the average time it takes for surfactant adsorption and τ_{coal} is the average time it takes until a droplet collides with another one). In the various regimes, the hydrodynamic constraints are the same for τ_{ads}. For example, in regime LV, $\tau_{coal} = \pi/8\varphi G$. Thus for all regimes, the ratio of τ_{ads}/τ_{coal} is given by [2]

$$\kappa \equiv \frac{\tau_{ads}}{\tau_{coal}} \propto \frac{\varphi \Gamma}{m_C d}, \tag{6.20}$$

where the proportionality factor would be at least of order 10. For example, for $\varphi = 0.1$, $\Gamma/m_C = 10^{-6}$ m and $d = 10^{-6}$ m (total surfactant concentration of the emulsion should then be about 5%), κ would be of the order of 1. For $\kappa \gg 1$, considerable coalescence is likely to occur, particularly at high φ. The coalescence rate would then markedly increase during emulsification, since both m_C and d become smaller during the process. If emulsification proceeds long enough, the droplet size distribution may then be the result of a steady state of simultaneous break-up and coalescence.

The effect of increase in φ can be summarised as follows [2]: (i) τ_{coal} is shorter and coalescence will be faster unless κ remains small; (ii) emulsion viscosity η_{em} increases, hence Re decreases; this implies a change of flow from turbulent to laminar (LV); (iii) in laminar flow, the effective η_C becomes higher. The presence of many droplets means that the local velocity gradients near a droplet will generally be higher than the overall value of G. Consequently, the local shear stress ηG does increase with increase of φ, which is as if η_C increases; (iv) in turbulent flow, increase in φ will induce turbulence depression leading to larger d; (v) if the mass ratio of surfactant to continuous phase is constant, an increase in φ gives a decrease in surfactant concentration; hence an increase in γ_{eq}, increase in κ, increase in d produced by increase in coalescence rate. If the mass ratio of surfactant to disperse phase is kept constant, the above mentioned changes are reversed, unless $\kappa \ll 1$.

It is clear from the above discussion that general conclusions cannot be drawn, since several of the above mentioned mechanisms may come into play. Using a high pressure homogenizer, Walstra [2] compared the values of d with various φ values up to 0.4 at constant initial m_C, regime TI probably changing to TV at higher φ. With increasing φ (> 0.1), the resulting d increased, and the dependence on homogeniser pressure p_H (Fig. 6.9). This points to increased coalescence (effects i and v).

Fig. 6.9 shows a comparison of the average droplet diameter vs power consumption using different emulsifying machines. It can be seen that the smallest droplet diameters were obtained when using the high pressure homogenizers.

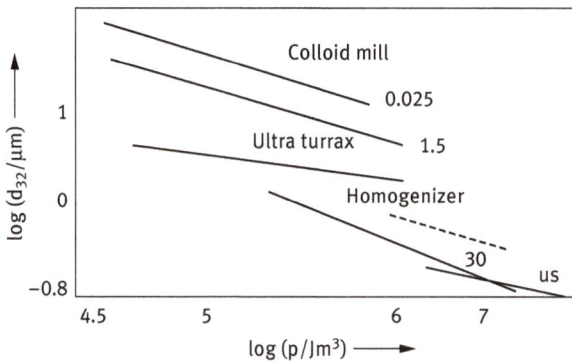

Fig. 6.9: Average droplet diameters obtained in various emulsifying machines as a function of energy consumption p. The number near the curves denote the viscosity ratio λ. The results for the homogenizer are for $\varphi = 0.04$ (solid line) and $\varphi = 0.3$ (broken line); us means ultrasonic generator.

References

[1] T. F. Tadros and B. Vincent, in: P. Becher (ed.), *Encyclopedia of Emulsion Technology*, Marcel Dekker, New York, 1983.
[2] P. Walstra and P. E. A. Smolders, in: B. P. Binks (ed.), *Modern Aspects of Emulsions*, The Royal Society of Chemistry, Cambridge, 1998.
[3] Tharwat Tadros, *Applied Surfactants*, Wiley-VCH, Germany, 2005.
[4] Tharwat Tadros, Emulsion Formation Stability and Rheology, in: Tharwat Tadros (ed.), *Emulsion Formation and Stability*, Ch. 1, Wiley-VCH, Germany, 2013.
[5] H. A. Stone, *Ann. Rev. Fluid Mech.* **226** (1994), 95.

[6] J. A. Wierenga, F. ven Dieren, J. J. M. Janssen and W. G. M. Agterof, *Trans. Inst. Chem. Eng.* **74-A** (1996), 554.

[7] V. G. Levich, *Physicochemical Hydrodynamics*, Prentic-Hall, Englewood Cliffs, 1962.

[8] J. T. Davis, *Turbulent Phenomena*, Academic Press, London, 1972.

7 Selection of emulsifiers

7.1 Introduction

Several surfactants and their mixtures are used for the preparation and stabilisation of oil-in-water (O/W) and water-in-oil (W/O) emulsions. A summary of the most commonly used surfactants is given below [1, 2].

Anionic Surfactants

- Carboxylates: $C_nH_{2n+1}COO^-X^+$
- Sulphates: $C_nH_{2n+1}OSO_3^-X^+$
- Sulphonates: $C_nH_{2n+1}SO_3^-X^+$
- Phosphates: $C_nH_{2n+1}OPO(OH)O^-X^+$

with n being the range 8–16 atoms and the counter-ion X^+ being usually Na^+.

Several other anionic surfactants are commercially available, such as sulphosuccinates, isethionates (esters of isothionic acid with the general formula $RCOOCH_2-CH_2-SO_3Na$) and taurates (derivatives of methyl taurine with the general formula $RCON(R')CH_2-CH_2-SO_3Na$), sarchosinates (with the general formula $RCON(R')COO$ Na), and these are sometimes used for special applications.

Cationic surfactants

The most common cationic surfactants are the quaternary ammonium compounds with the general formula $R'R''R'''R''''N^+X^-$, where X^- is usually chloride ion and R represents alkyl groups. These quaternaries are made by reacting an appropriate tertiary amine with an organic halide or organic sulphate. Another class of cationics are those based on pyridinium salts such as lauryl pyridinium chloride.

Amphoteric (zwitter-ionic) surfactants

These are surfactants containing both cationic and anionic groups. The most common amphoterics are the N-alkyl betaines, which are derivatives of trimethyl glycine $(CH_3)_3NCH_2COOH$ (described as betaine). An example of betaine surfactant is lauryl amido propyl dimethyl betaine $C_{12}H_{25}CON(CH_3)_2CH_2COOH$. These alkyl betaines are sometimes described as alkyl dimethyl glycinates. The main characteristics of amphoteric surfactants is their dependence on the pH of the solution in which they are dissolved. In acid pH solutions, the molecule acquires a positive charge and behaves like a cationic, whereas in alkaline pH solutions they become negatively charged and behave like an anionic. A specific pH can be defined at which both ionic groups show equal ionization (the isoelectric point of the molecule).

Nonionic Surfactants

The most common nonionic surfactants are those based on ethylene oxide, referred to as ethoxylated surfactants. Several classes can be distinguished: alcohol ethoxylates, alkyl phenol ethoxylates, fatty acid ethoxylates, monoalkaolamide ethoxylates, sorbitan ester ethoxylates, fatty amine ethoxylates and ethylene oxide–propylene oxide copolymers (sometimes referred to as polymeric surfactants). Another important class of nonionics are the multihydroxy products such as glycol esters, glycerol (and polyglycerol) esters, glucosides (and polyglucosides) and sucrose esters. Amine oxides and sulphinyl surfactants represent nonionics with a small head group.

The fatty acid esters of sorbitan (generally referred to as Spans, an Atlas commercial trade name) and their ethoxylated derivatives (generally referred to as Tweens) are perhaps one of the most commonly used emulsifiers. The sorbitan esters are produced by reaction of sorbitol with a fatty acid at a high temperature ($> 200\,°C$). The sorbitol dehydrates to 1,4-sorbitan, and then esterification takes place. If one mole of fatty acid is reacted with one mole of sorbitol, one obtains a mono-ester (some di-ester is also produced as a by-product). Thus, sorbitan mono-ester has the following general formula,

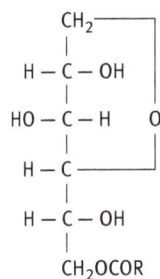

$$
\begin{array}{l}
CH_2{-} \\
| \\
H-C-OH \\
| \\
HO-C-H \quad\quad O \\
| \\
H-C{-} \\
| \\
H-C-OH \\
| \\
CH_2OCOR
\end{array}
$$

The free OH groups in the molecule can be esterified, producing di- and tri-esters. Several products are available depending on the nature of the alkyl group of the acid and whether the product is a mono-, di- or tri-ester. Some examples are given below:
- Sorbitan monolaurate – Span 20
- Sorbitan monopalmitate – Span 40
- Sorbitan monostearate – Span 60
- Sorbitan mono-oleate – Span 80

- Sorbitan tristearate – Span 65
- Sorbitan trioleate – Span 85

The ethoxylated derivatives of Spans (Tweens) are produced by reaction of ethylene oxide on any hydroxyl group remaining on the sorbitan ester group. Alternatively, the sorbitol is first ethoxylated and then esterified. However, the final product has differ- ent surfactant properties to the Tweens. Some examples of Tween surfactants are as follows:

- Polyoxyethylene (20) sorbitan monolaurate – Tween 20
- Polyoxyethylene (20) sorbitan monopalmitate – Tween 40
- Polyoxyethylene (20) sorbitan monostearate – Tween 60
- Polyoxyethylene (20) sorbitan mono-oleate – Tween 80
- Polyoxyethylene (20) sorbitan tristearate – Tween 65
- Polyoxyethylene (20) sorbitan tri-oleate – Tween 85

The sorbitan esters are insoluble in water, but soluble in most organic solvents (low HLB number surfactants; see below). The ethoxylated products are generally soluble in number, and they have relatively high HLB numbers. One of the main advantages of the sorbitan esters and their ethoxylated derivatives is their approval as food additives. They are also widely used in cosmetics and some pharmaceutical emulsions.

Polymeric surfactants

The simplest type of a polymeric surfactant is a homopolymer, which is formed from the same repeating units, such as poly(ethylene oxide) or poly(vinyl pyrrolidone). These homopolymers have little surface activity at the o/w interface, since the ho- mopolymer segments (ethylene oxide or vinylpyrrolidone) are highly water soluble and have little affinity to the interface. Clearly, homopolymers are not the most suit- able emulsifiers. A small variant is to use polymers that contain specific groups that have high affinity to the surface. This is exemplified by partially hydrolysed poly(vinyl acetate) (PVAc), technically referred to as poly(vinyl alcohol) (PVA). These partially hydrolysed PVA molecules exhibit surface activity at the O/W interface. The most con- venient polymeric surfactants are those of the block and graft copolymer type. A block copolymer is a linear arrangement of blocks of variable monomer composition. The nomenclature for a diblock is poly-A-block-poly-B and for a triblock is poly-A-block- poly-B-poly-A. One of the most widely used triblock polymeric surfactants are the "Pluronics" (BASF, Germany), which consists of two poly-A blocks of poly(ethylene oxide) (PEO) and one block of poly(propylene oxide) (PPO). Several chain lengths of PEO and PPO are available. Two types may be distinguished: those prepared by reaction of polyoxypropylene glycol (dinfunctional) with EO or mixed EO/PO, giving block copolymers with the structure

$HO(CH_2CH_2O)_n-(CH_2CHO)_m-(CH_2CH_2)_nOH$, abbreviated $(EO)_n(PO)_m(EO)_n$
$$\quad\quad\quad\quad\quad\quad\quad | $$
$$\quad\quad\quad\quad\quad\quad CH_3$$

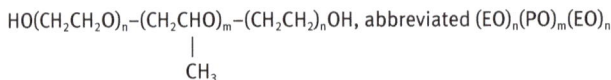

Various molecules are available, where n and m are varied systematically. The second type of EO/PO copolymers are prepared by reaction of polyethylene glycol (difunctional) with PO or mixed EO/PO. These will have the structure $(PO)_n(EO)_m(PO)_n$, and they are referred to as reverse pluronics. These polymeric triblocks can be applied as emulsifiers, whereby the assumption is made that the hydrophobic PPO chain resides at the hydrophobic oil surface, leaving the two PEO chains dangling in aqueous solution and hence providing steric repulsion.

It is clear from the above list that a vast number of surfactants can be used as emulsifiers, which makes the choice rather difficult. A number of semi-empirical methods have been applied for surfactant selection. The most commonly used method is based on the hydrophilic-lipophile balance (HLB) concept. A closely related procedure is based on measurement of the phase inversion temperature (PIT) which is particularly useful for selection of non-ionic surfactants based on polyethylene oxide. A more quantitative method for surfactant selection is the cohesive energy ratio (CER) concept, which considers the interaction between the lipophilic chain with the oil and the hydrophilic chain with the aqueous phase. Another method for surfactant selection is based on the critical packing parameter (CPP) concept that considers the surfactant geometry and its optimum fitting at the O/W interface. A description of all these methods is given below.

7.2 The hydrophilic-lipophile balance (HLB) concept

The hydrophilic-lipophilic balance (HLB number) is a semi-empirical scale for selecting surfactants developed by Griffin [3]. This scale is based on the relative percentage of hydrophilic to lipophilic (hydrophobic) groups in the surfactant molecule(s). For an O/W emulsion droplet the hydrophobic chain resides in the oil phase, whereas the hydrophilic head group resides in the aqueous phase. For a W/O emulsion droplet, the hydrophilic group(s) reside in the water droplet, whereas the lipophilic groups reside in the hydrocarbon phase.

Tab. 7.1 gives a guide to the selection of surfactants for a particular application. The HLB number depends on the nature of the oil. As an illustration, Tab. 7.2 gives the required HLB numbers to emulsify various oils. Examples of HLB numbers of a list of surfactants are given in Tab. 7.3.

The relative importance of the hydrophilic and lipophilic groups was first recognised when using mixtures of surfactants containing varying proportions of a low and high HLB number. The efficiency of any combination (as judged by phase separation) was found to pass a maximum when the blend contained a particular proportion of the

Tab. 7.1: Summary of HLB ranges and their applications.

HLB range	Application
3–6	W/O emulsifier
7–9	Wetting agent
8–18	O/W emulsifier
13–15	Detergent
15–18	Solubiliser

Tab. 7.2: Required HLB numbers to emulsify various oils.

Oil	W/O emulsion	O/W emulsion
Paraffin oil	4	10
Beeswax	5	9
Linolin, anhydrous	8	12
Cyclohexane	–	15
Toluene	–	15
Silicone oil (volatile)	–	7–8
Isopropyl myristate	–	11–12
Isohexadecyl alcohol	–	11–12
Castor oil	–	14

Tab. 7.3: HLB numbers of some surfactants.

Surfactant	Chemical name	HLB
Span 85	Sorbitan trioleate	1.8
Span 80	Sorbitan monooleate	4.3
Brij 72	Ethoxylated (2 mol ethylene oxide) stearyl alcohol	4.9
Triton X-35	Ethoxylated octylphenol	7.8
Tween 85	Ethoxylated (20 mol ethylene oxide) sorbitan trioleate	11.0
Tween 80	Ethoxylated (20 mol ethylene oxide) sorbitan monooleate	15.0

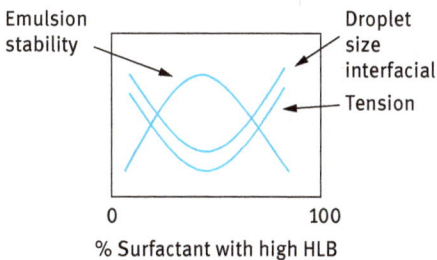

Fig. 7.1: Variation of emulsion stability, droplet size and interfacial tension with % surfactant with high HLB number.

surfactant with the higher HLB number. This is illustrated in Fig. 7.1, which shows the variation of emulsion stability, droplet size and interfacial tension with % surfactant with high HLB number.

The average HLB number may be calculated from additivity:

$$HLB = x_1 HLB_1 + x_2 HLB_2 , \tag{7.1}$$

where x_1 and x_2 are the weight fractions of the two surfactants with HLB_1 and HLB_2.

Griffin [3] developed simple equations for calculation of the HLB number of relatively simple nonionic surfactants. For a polyhydroxy fatty acid ester,

$$HLB = 20\left(1 - \frac{S}{A}\right), \tag{7.2}$$

where S is the saponification number of the ester and A is the acid number. For a glyceryl monostearate, S = 161 and A = 198; the HLB is 3.8 (suitable for W/O emulsion).

For a simple alcohol ethoxylate, the HLB number can be calculated from the weight percent of ethylene oxide (E) and Polyhydric alcohol (P):

$$HLB = \frac{E + P}{5}. \tag{7.3}$$

If the surfactant contains PEO as the only hydrophilic group, the contribution from one OH group can be neglected

$$HLB = \frac{E}{5}. \tag{7.4}$$

For a nonionic surfactant $C_{12}H_{25}-O-(CH_2-CH_2-O)_6$, the HLB is 12 (suitable for O/W emulsion).

The above simple equations cannot be used for surfactants containing propylene oxide or butylene oxide. They cannot be also applied for ionic surfactants. Davies [4, 5] devised a method for calculating the HLB number for surfactants from their chemical formulae, using empirically determined group numbers. A group number is assigned to various component groups. A summary of the group numbers for some surfactants is given in Tab. 7.4.

The HLB is given by the following empirical equation,

$$HLB = 7 + \sum(\text{hydrophilic group numbers}) - \sum(\text{lipophilic group numbers}) \tag{7.5}$$

Davies has shown that the agreement between HLB numbers calculated from the above equation and those determined experimentally is quite satisfactory.

Various other procedures were developed to obtain a rough estimate of the HLB number. Griffin found good correlation between the cloud point of 5 % solution of various ethoxylated surfactants and their HLB number.

Davies [4, 5] attempted to relate the HLB values to the selective coalescence rates of emulsions. Such correlations were not realised, since it was found that the emulsion stability and even its type depends to a large extent on the method of dispersing the

Tab. 7.4: HLB group numbers.

	Group number
Hydrophilic	
$-SO_4Na^+$	38.7
$-COO-$	21.2
$-COONa$	19.1
N (tertiary amine)	9.4
Ester (sorbitan ring)	6.8
$-O-$	1.3
CH– (sorbitan ring)	0.5
Lipophilic	
$(-CH-), (-CH_2-), CH_3$	0.475
Derived	
$-CH_2-CH_2-O$	0.33
$-CH_2-CHCH_3-O-$	0.11

oil into the water, and vice versa. At best the HLB number can only be used as a guide for selecting optimum compositions of emulsifying agents.

One may take any pair of emulsifying agents which fall at opposite ends of the HLB scale, e.g. Tween 80 (sorbitan monooleate with 20 mol EO, HLB = 15) and Span 80 (sorbitan monooleate, HLB = 5) using them in various proportions to cover a wide range of HLB numbers. The emulsions should be prepared in the same way, with a few percent of the emulsifying blend. For example, a 20 % O/W emulsion is prepared by using 4 % emulsifier blend (20 % with respect to oil) and 76 % water. The stability of the emulsions is then assessed at each HLB number from the rate of coalescence, or qualitatively by measuring the rate of oil separation. In this way one may be able to find the optimum HLB number for a given oil. For example with a given oil, the optimum HLB number is found to be 10.3. The latter can be deterimned more exactly by using mixtures of surfactants with narrower HLB range, say between 9.5 and 11. Having found the most effective HLB value, various other surfactant pairs are compared at this HLB value, to find the most effective pair. This is illustrated in Fig. 7.2, which shows schematically the difference between three chemical classes of surfactants. Although the different classes give a stable emulsion at HLB 12, mixture A gives the best emulsion stability.

The HLB value of a given magnitude can be obtained by mixing emulsifiers of different chemical types. The "correct" chemical type is as important as the "correct" HLB number. This is illustrated in Fig. 7.3, which shows that an emulsifier with unsaturated alky chain such as oleate (ethoxylated sorbitan monooleate, Tween 80) is more suitable for emulsifying an unsaturated oil [6]. An emulsifier with saturated alkyl chain (stearate in Tween 60) is better for emulsifying a saturated oil).

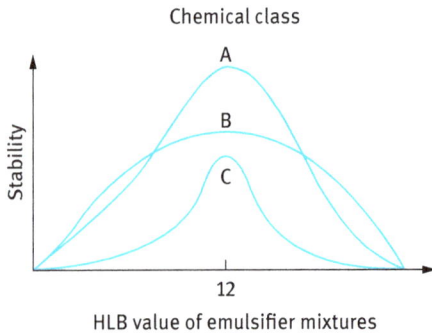

Fig. 7.2: Stabilisation of emulsion by different classes of surfactants as a function of HLB.

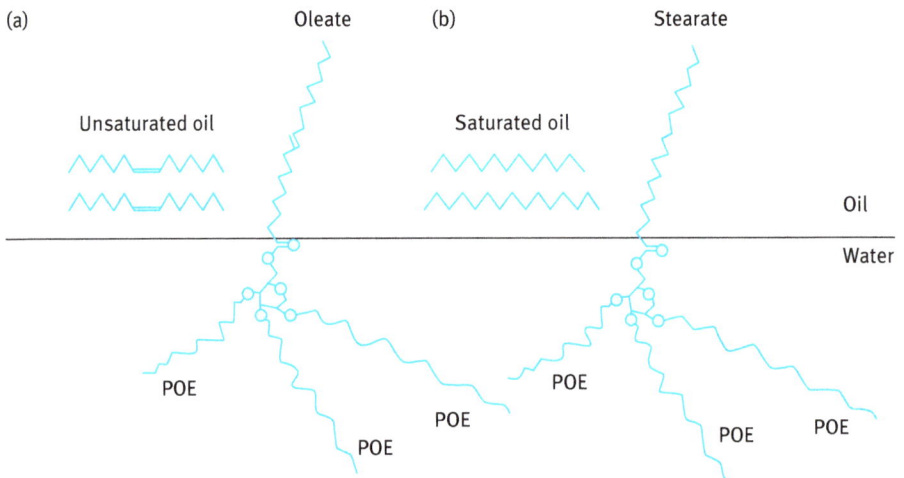

Fig. 7.3: Selection of Tween type to correspond to the type of the oil to be emulsified.

Various procedures have been developed to determine the HLB of different surfactants. Griffin [3] found a correlation between the HLB and the cloud points of 5 % aqueous solution of ethoxylated surfactants, as illustrated in Fig. 7.4. A titration procedure was developed [7] for estimating the HLB number. In this method, a 1 % solution of surfactant in benzene plus dioxane is titrated with distilled water at constant temperature until a permanent turbidity appears. They found a good linear relationship between the HLB number and the water titration value for polyhydric alcohol esters as shown in Fig. 7.5. However, the slope of the line depends on the class of material used.

Gas liquid chromatography (GLC) could also be used to determine the HLB number [7]. Since in GLC the efficiency of separation depends on the polarity of the substrate with respect to the components of the mixture, it should be possible to determine the HLB directly by using the surfactant as the substrate and passing an oil phase down the column. Thus, when a 50 : 50 mixture of ethanol and hexane is passed

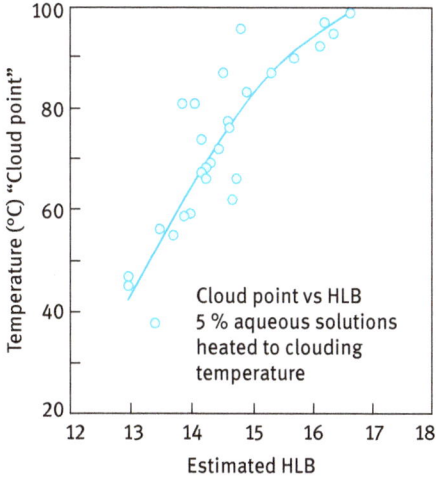

Fig. 7.4: Relationship between cloud point and HLB.

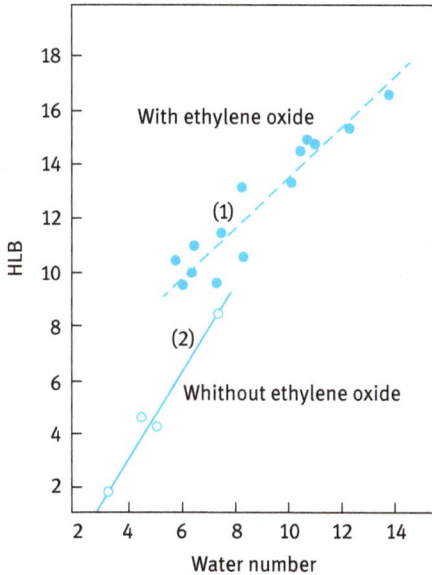

Fig. 7.5: Correlation of HLB with water number.

down a column of a simple non-ionic surfactant, such as sorbitan fatty acid esters and polyoxyethylated sorbitan fatty acid esters, two well-defined peaks, corresponding to hexane (which appears firs) and ethanol, appear on the chromatograms. A good correlation was found between the retention time ratio R_t (ethanol/hexane) and the HLB value. This is illustrated in Fig. 7.6. Statistical analysis of the data gave the following empirical relationship between R_t and HLB:

$$HLB = 8.55R_t - 6.36, \tag{7.6}$$

where

$$R_t = \frac{R_t^{ETOH}}{R_t^{hexane}} .$$

(7.7)

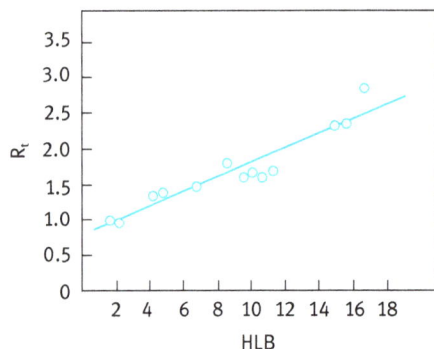

Fig. 7.6: Correlation between retention time and HLB of sorbitan fatty acid esters and polyoxyethylated fatty acid esters.

7.3 The phase inversion temperature (PIT) concept

Shinoda and coworkers [8, 9] found that many O/W emulsions stabilised with non-ionic surfactants undergo a process of inversion at a critical temperature (PIT). The PIT can be determined by following the emulsion conductivity (a small amount of electrolyte is added to increase the sensitivity) as a function of the temperature, as illustrated in Fig. 7.7. The conductivity of the O/W emulsion increases with the increase of temperature until the PIT is reached, above which there will be a rapid reduction in conductivity (W/O emulsion is formed). Shinoda and coworkers [8, 9] found that the PIT is influenced by the HLB number of the surfactant [10], as shown in Fig. 7.8. For any given oil, the PIT increases with increase of the HLB number. The size of the emulsion droplets was found to depend on the temperature and HLB number of the emulsifiers. The droplets are less stable towards coalescence close to the PIT. However, a stable system may be produced by rapid cooling of the emulsion. Relatively stable O/W emulsions were obtained when the PIT of the system was 20–65 °C higher than the storage temperature. Emulsions prepared at a temperature just below the PIT followed by rapid cooling generally have smaller droplet sizes. This can be understood if one considers the change of interfacial tension with temperature, as illustrated in Fig. 7.9. The interfacial tension decreases with increase of temperature reaching a minimum close to the PIT, after which it increases.

Thus, the droplets prepared close to the PIT are smaller than those prepared at lower temperatures. These droplets are relatively unstable towards coalescence near the PIT, but by rapid cooling of the emulsion one can retain the smaller size. This procedure may be applied to prepare mini (nano) emulsions.

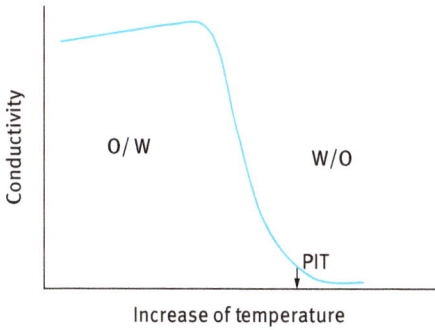

Fig. 7.7: Variation of conductivity with temperature for an O/W emulsion.

The optimum stability of the emulsion was found to be relatively insensitive to changes
in the HLB value or the PIT of the emulsifier, but instability was very sensitive to the
PIT of the system. It is essential, therefore to measure the PIT of the emulsion as a
whole (with all other ingredients).

At a given HLB value, stability of the emulsions against coalescence increases
markedly as the molar mass of both the hydrophilic and lipophilic components increases. The enhanced stability using high molecular weight surfactants (polymeric
surfactants) can be understood from a consideration of the steric repulsion, which
produces more stable films produced using macromolecular surfactants resist thinning and disruption thus reducing the possibility of coalescence. The emulsions

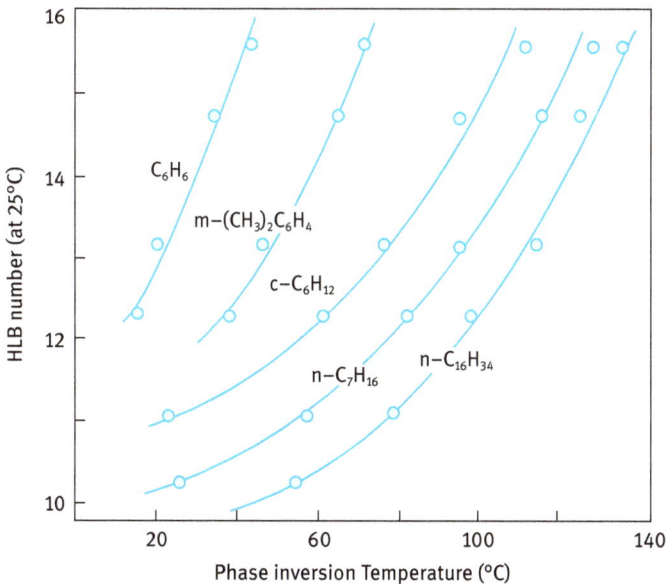

Fig. 7.8: Correlation between HLB number and PIT for various O/W (1 : 1) emulsions
stabilised with non-ionic surfactants (1.5 wt %).

showed maximum stability when the distribution of the PEO chains was broad. The cloud point is lower, but the PIT is higher than in the corresponding case for narrow size distributions. The PIT and HLB number are directly related parameters.

Adding electrolytes reduces the PIT, and hence an emulsifier with a higher PIT value is required when preparing emulsions in the presence of electrolytes. Electrolytes cause dehydration of the PEO chains, and in effect this reduces the cloud point of the non-ionic surfactant. One needs to compensate for this effect by using a surfactant with higher HLB. The optimum PIT of the emulsifier is fixed if the storage temperature is fixed.

In view of the above correlation between PIT and HLB and the possible dependence of the kinetics of droplet coalescence on the HLB number, Sherman and coworkers suggested the use of PIT measurements as a rapid method for assessing emulsion stability. However, one should be careful in using such methods for assessment of the long term stability, since the correlations were based on a very limited number of surfactants and oils.

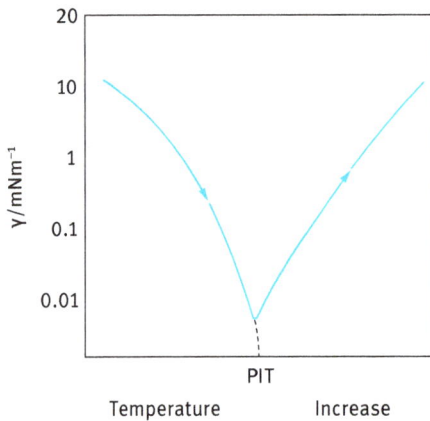

Fig. 7.9: Variation of interfacial tension with temperature increase for an O/W emulsion.

Measurement of the PIT can at best be used as a guide for preparation of stable emulsions. Assessment of the stability should be evaluated by following the droplet size distribution as a function of time using a Coulter counter or light diffraction techniques. Following the rheology of the emulsion as a function of time and temperature may also be used for assessment of the stability against coalescence [11]. Care should be taken in analysing the rheological results. Coalescence results in an increase in the droplet size, and this is usually followed by a reduction in the viscosity of the emulsion. This trend is only observed if the coalescence is not accompanied by flocculation of the emulsion droplets (which results in an increase in the viscosity). Ostwald ripening can also complicate the analysis of the rheological data.

7.4 The cohesive energy ratio (CER) concept

Beerbower and Hills [12] considered the dispersing tendency on the oil and water interfaces of the surfactant or emulsifier in terms of the ratio of the cohesive energies of the mixtures of oil with the lipophilic portion of the surfactant and the water with the hydrophilic portion. They used the Winsor R_0 concept, which is the ratio of the intermolecular attraction of oil molecules (O) and lipophilic portion of surfactant (L), C_{LO}, to that of water (W) and hydrophilic portion (H), C_{HW}:

$$R_0 = \frac{C_{LO}}{C_{HW}} \; . \tag{7.8}$$

Several interaction parameters may be identified at the oil and water sides of the interface. One can identify at least nine interaction parameters as schematically represented in Fig. 7.10.

C_{LL}, C_{OO}, C_{LO} (at oil side)

C_{HH}, C_{WW}, C_{HW} (at water side)

C_{LW}, C_{HO}, C_{LH} (at the interface)

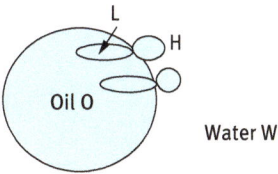

Fig. 7.10: The cohesive energy ratio concept.

In the absence of emulsifier, there will be only three interaction parameters: C_{OO}, C_{WW}, C_{OW}; if $C_{OW} \ll C_{WW}$, the emulsion breaks.

The above interaction parameters may be related to the Hildebrand solubility parameter δ [13] (at the oil side of the interface) and the Hansen [14] non-polar, hydrogen bonding and polar contributions to δ at the water side of the interface.

The solubility parameter of any component is related to its heat of vaporization ΔH by the expression

$$\delta^2 = \frac{\Delta H - RT}{V_m} \, , \tag{7.9}$$

where V_m is the molar volume.

Hansen [14] considered δ (at the water side of the interface) to consist of three main contributions: a dispersion contribution, δ_d, a polar contribution, δ_p and a hydrogen bonding contribution, δ_h. These contributions have different weighting factors:

$$\delta^2 = \delta_d^2 + 0.25\delta_p^2 + 0.25\delta_h^2 . \tag{7.10}$$

Beerbower and Hills [12] used the following expression for the HLB number:

$$HLB = 20 \left(\frac{M_H}{M_L + M_H} \right) = 20 \left(\frac{V_H \rho_H}{V_L \rho_L + V_H \rho_H} \right),$$ (7.11)

where M_H and M_L are the molecular weights of the hydrophilic and lipophilic portions of the surfactants. V_L and V_H are their corresponding molar volumes, whereas ρ_H and ρ_L are the densities, respectively.

The cohesive energy ratio was originally defined by Winsor, equation (7.8).

When $C_{LO} > C_{HW}$, $R > 1$ and a W/O emulsion forms. If $C_{LO} < C_{HW}$, $R < 1$ and an O/W emulsion forms. If $C_{LO} = C_{HW}$, $R = 1$ and a planer system results: this denotes the inversion point.

Ro can be related to V_L, δ_L and V_H, δ_H by the expression

$$R_o = \frac{V_L \delta_L^2}{V_H \delta_H^2}.$$ (7.12)

Using equation (7.11),

$$R_o = \frac{V_L (\delta_d^2 + 0.25\delta_p^2 + 0.25\delta_h^2)_L}{V_h (\delta_d^2 + 0.25\delta_p^2 + 0.25\delta_h^2)_H}.$$ (7.13)

Combining equations (7.11) and (7.13), one obtains the following general expression for the cohesive energy ratio:

$$R_o = \left(\frac{20}{HLB} - 1 \right) \frac{\rho_h (\delta_d^2 + 025\delta_p^2 + 0.25\delta_h^2)_L}{\rho_L (\delta_d^2 + 0.25\delta_p^2 + 0.25\delta_p^2)_L}.$$ (7.14)

For O/W system, HLB = 12–15 and R_o = 0.58–0.29 ($R_o < 1$). For W/O system, HLB = 5–6 and R_o = 2.3–1.9 ($R_o > 1$). For a planer system, HLB = 8–10 and R_o = 1.25–0.85 ($R_o \sim 1$).

The R_o equation combines both the HLB and cohesive energy densities; it gives a more quantitative estimate of emulsifier selection. R_o considers HLB, molar volume and chemical match. The success of this approach depends on the availability of data on the solubility parameters of the various surfactant portions. Some values are tabulated in the book by Barton [15].

7.5 The critical packing parameter (CPP) for emulsion selection

The critical packing parameter (CPP) is a geometric expression relating the hydrocarbon chain volume (v) and length (l) and the interfacial area occupied by the head group (a_o) [16]:

$$CPP = \frac{v}{l_c a_o},$$ (7.15)

where a_o is the optimal surface per head group and l_c is the critical chain length.

Regardless of the shape of any aggregated structure (spherical or cylindrical micelle or a bilayer), no point within the structure can be farther from the hydrocarbon-water surface than l_c. The critical chain length l_c is roughly equal to but less than the fully extended length of the alkyl chain.

The above concept can be applied to predict the shape of an aggregated structure. Consider a spherical micelle with radius r and aggregation number n; the volume of the micelle is given by

$$\left(\frac{4}{3}\right)\pi r^3 = nv,$$ (7.16)

where v is the volume of a surfactant molecule.

The area of the micelle is given by

$$4\pi r^2 = na_o,$$ (7.17)

where a_o is the area per surfactant head group.

Combining equations (7.16) and (7.17),

$$a_o = \frac{3v}{r}$$ (7.18)

The cross sectional area of the hydrocarbon chain a is given by the ratio of its volume to its extended length l_c:

$$a = \frac{v}{l_c}.$$ (7.19)

From (7.18) and (7.19),

$$P = \frac{a}{a_o} = \left(\frac{1}{3}\right)\left(\frac{r}{l_c}\right).$$ (7.20)

Since $r < l_c$, then CPP $\leq \frac{1}{3}$.

For a cylindrical micelle with length d and radius r,

$$\text{Volume of the micelle} = \pi r^2 d = nv,$$ (7.21)
$$\text{Area of the micelle} = 2\pi rd = na_o.$$ (7.22)

Combining equations (7.21) and (7.22),

$$a_o = \frac{2v}{r}$$ (7.23)

$$a = \frac{v}{l_c}$$ (7.24)

$$\text{CPP} = \frac{a}{a_o} = \left(\frac{1}{2}\right)\left(\frac{r}{l_c}\right).$$ (7.25)

Since $r < l_c$, then $\frac{1}{3} < \text{CPP} \leq \frac{1}{2}$.

For vesicles (liposomes) $1 > \text{CPP} \geq \frac{2}{3}$, and for lamellar micelles $P \sim 1$. For inverse micelles CPP > 1. A summary of the various shapes of micelles and their CPP is given in Tab. 7.5.

Surfactants that make spherical micelles with the above packing constraints, i.e. CPP $\leq \frac{1}{3}$, are more suitable for O/W emulsions. Surfactants with CPP > 1, i.e. forming inverted micelles, are suitable for formation of W/O emulsions.

Tab. 7.5: CPP and shape of micelles.

Lipid	Critical packing parameter $v/a_0 l_c$	Critical packing shape	Structures formed
Single-chained lipids (surfactants) with large head-group areas: – *SDS in low salt*	$< 1/3$	Cone	Spherical micelles
Single-chained lipids with small head-group areas: – *SDS and CTAB in high salt* – *nonionic lipids*	$1/3–1/2$	Truncated cone	Cylindrical micelles
Double-chained lipids with large head-group areas, fluid chains: – *Phosphatidyl choline (lecithin)* – *phosphatidyl serine* – *phosphatidyl glycerol* – *phosphatidyl inositol* – *phosphatidic acid* – *sphingomyelin, DGDG*[a] – *dihexadecyl phosphate* – *dialkyl dimethyl ammonium* – *salts*	$1/2–1$	Truncated one	Flexible bilayers, vesicles
Double-chained lipids with small head-group areas, anionic lipids in high salt, saturated frozen chains: – *phosphatidyl ethanaiamine* – *phosphatidyl serine + Ca*$^{2+}$	~ 1	Cylinder	Planar bilayers
Double-chained lipids with small head-group areas, nonionic lipids, poly(cis) unsaturated chains, high T: – *unsat. phosphatidyl ethanolamine* – *cardiolipin + Ca*$^{2+}$ – *phosphatidic acid + Ca*$^{2+}$ – *cholesterol, MGDG*[b]	> 1	Inverted truncated cone or wedge	Inverted micelles

a DGDG: digalactosyl diglyceride, diglucosyldiglyceride
b MGDG: monogalactosyl diglyceride, monoglucosyl diglyceride

7.6 Stabilisation by solid particles (Pickering emulsions)

Very effective stabilisation against coalescence can be obtained by using finely divided solids as emulsifying agents [17]. These emulsions are sometimes referred to as Pickering emulsions. The type of emulsion produced depends on which phase preferentially wets the solid particles. If the particles are wetted more by the oil phase (hydrophobic particles such as gas black), a W/O emulsion is produced; where the aqueous phase preferentially wets the solid (such as bentonite), an O/W emulsion is produced. This is schematically illustrated in Fig. 7.11.

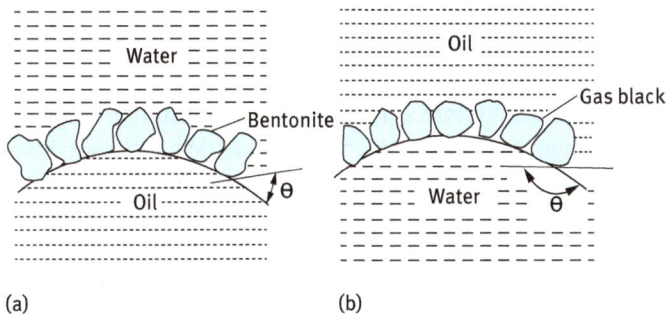

(a)   (b)

Fig. 7.11: Schematic representation of stabilisation of an emulsion with solid particles: (a) hydrophilic (bentonite) and (b) hydrophobic (gas black) particles.

The fine particles at the interface prevent both coalescence and flocculation (if the particles repel one another at the interface). Solid particles with a contact angle θ of 90° as illustrated in Fig. 7.12 form the most stable emulsions [6].

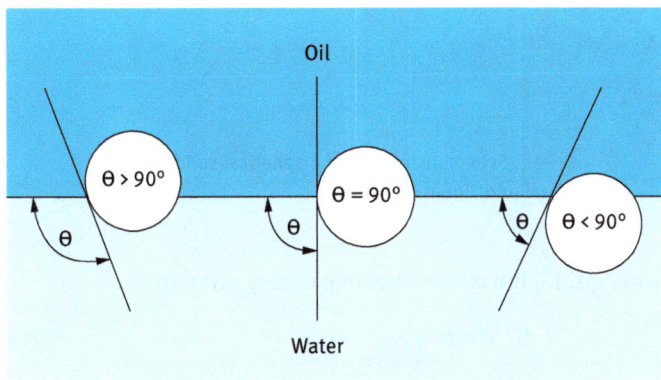

Fig. 7.12: Contact angle of solid particles at the O/W interface.

In the presence of surface active agents, O/W emulsions are produced when the contact angle is slightly less than 90°, whereas W/O are produced when the contact angle is slightly higher than 90°. Thus, the contact angle at the three phase (oil/water/solid) boundary is critical in controlling the stabilisation by given solid particles [7]. These particles have to be very small in comparison with the oil droplets. As mentioned above, an optimum contact angle is required to collect the particles at the liquid/liquid interface by capillary forces. If the particles are too hydrophilic (such as silica or alumina), they will pass into the aqueous phase too quickly, whereas if the particles are too hydrophobic (e.g. certain carbon blacks), then W/O emulsions are produced.

The mechanism of stabilisation of emulsion droplets by solid particles is far from being fully understood. Presumably the presence of solid particles at the liquid/liquid interface plays an important role in preventing the thinning of the liquid film between the droplets. For the solid particles to be effective they should form a continuous mono-particulate film. Contact angle hysteresis is probably important in preventing displacement of the meniscus. For that reason "rough" asymmetric particles such as bentonite clays are more effective than "smooth" spherical particles. The particles should also form a coherent film at the interface. This is achieved by capillary forces, which bring the particles close together at the interface. The smaller the radius of curvature of the meniscus between the particles, the larger the force of attraction; this explains the need for very fine solids [7].

For the solid particles to be located at the interface, the magnitude of the contact angles formed between the two liquid phases plays an essential role, as discussed above. It is necessary to consider the balance of forces for a solid sphere at the liquid/liquid interface. Let us for simplicity consider a simple model of spherical particles of radius r, having the same density as two liquids 1 and 2 (again of equal density), i.e. there are no effects from gravitational forces (Fig. 7.13).

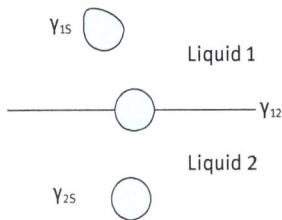

Fig. 7.13: Schematic picture of a spherical solid particle at the liquid/liquid interface.

If the particle is entirely in liquid 1, it has a surface free energy given by

$$G_1 = 4\pi r^2 \gamma_{1S} ,$$
(7.26)

where γ_{1S} is the interfacial tension between the solid particle and liquid 1. Similarly, if the particle is completely immersed in liquid 2, it has a surface free energy G_2 given by

$$G_2 = 4\pi r^2 \gamma_{2S} .$$
(7.27)

If the particle stays at the interface, it will be partly immersed in liquid 1 and partly in liquid 2. If ΔA_{1S} and ΔA_{2S} are the areas of the sphere immersed in liquid 1 and 2, respectively, then the surface free energy at the interface is given by

$$G = \Delta A_{1S}\gamma_{1S} + \Delta A_{2S}\gamma_{2S}. \qquad (7.28)$$

When the sphere is located at the interface, it displaces an area ΔA_{12} of the interface with an interfacial tension γ_{12}. At equilibrium, the total surface free energy must be at a minimum:

$$dG = \Delta A_{1S}\gamma_{1S} + \Delta A_{2S}\gamma_{2S} + \Delta A_{12}\gamma_{12} = 0. \qquad (7.29)$$

The areas ΔA_{1S}, ΔA_{2S}, ΔA_{12} can be calculated from simple geometry as shown in Fig. 7.14

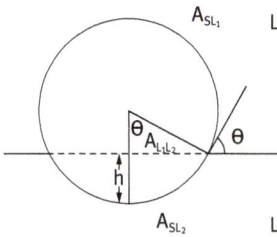

Fig. 7.14: Solid sphere at the liquid/liquid interface.

Thus,

$$2\pi r dh\gamma_{2S} + (-2\pi r dh)\gamma_{1S} - \pi(2r - 2h)\, dh\gamma_{12} = 0. \qquad (7.30)$$

This simplifies to

$$\gamma_{1S} - \gamma_{2S} = \left(1 - \frac{h}{r}\right)\gamma_{12}. \qquad (7.31)$$

Since

$$\cos\theta = 1 - \frac{h}{r}, \qquad (7.32)$$

then

$$\gamma_{1S} - \gamma_{2S} = \gamma_{12}\cos\theta, \qquad (7.33)$$

which is essentially the Young's equation for contact angle equilibrium [1].

Thus, a particle will seek a stable position at the interface such that the angle θ becomes the equilibrium contact angle. Due to the influence of gravity, there will be a slight displacement from this position until the net force due to the interfacial tension effects just equals the gravity force. Particles forming a finite contact angle at the liquid/liquid interface remain there, whereas those completely wetted by either of the two liquids will be displaced into the liquid phase.

The barrier produced by the solid particles which prevents coalescence can be understood if one considers the free-energy changes involved when a particle at equilibrium at the interface is displaced to either of the liquid phases. If the particle is removed into the water phase, the area of the O/W interface is increased by $\pi r^2 \sin^2 \theta$, and an area $2\pi r^2 (1 - \cos \theta)$ of the solid would be transferred from oil to water. The surface free energy increase would be given by

$$\Delta G = \pi r^2 \sin \theta \gamma_{OW} + 2\pi r^2 (1 + \cos \theta)(\gamma_{WS} - \gamma_{OS}). \tag{7.34}$$

Since

$$\gamma_{OS} \cos \theta = \gamma_{WS} - \gamma_{OS}, \tag{7.35}$$

then

$$\begin{aligned}
\Delta G &= \pi r^2 \sin \theta \gamma_{OW} + 2\pi r^2 (1 + \cos \theta)(\gamma_{WS} - \gamma_{OS}) \\
&= \pi r^2 \gamma_{OW} [\sin^2 \theta + 2 \cos \theta (1 + \cos \theta)] \\
&= \pi r^2 \gamma_{OW} (1 + \cos \theta)^2
\end{aligned} \tag{7.36}$$

Similarly the free-energy increase on transfer of the sphere from its equilibrium position to the oil phase would be

$$\begin{aligned}
\Delta G &= \pi r^2 \gamma_{OW} [\sin^2 \theta - 2 \cos \theta (1 + \cos \theta)] \\
&= \pi r^2 \gamma_{OW} (1 - \cos \theta)^2.
\end{aligned} \tag{7.37}$$

The difference, being the energy increase on transfer from oil to water, is

$$4\pi r^2 \gamma_{OS} \cos \theta = 4\pi r^2 (\gamma_{WS} - \gamma_{OS}). \tag{7.38}$$

Thus, if $\theta = 90°$, i.e. the equilibrium position for the particle is half in the water phase and half in the oil phase (see Fig. 7.12), the free energy necessary to take the sphere from the interface to either liquid is the same and $\gamma_{WS} - \gamma_{OS} = 0$; the free energy change required is provided entirely by elimination of some O/W interface. Thus, even if the spheres form a close packed layer with $\theta = 90°$, not all the O/W interface can be eliminated. The maximum proportion which could be eliminated is $\frac{\pi}{2} \cdot 3^{1/2} = 0.91$. Thus, although the movement of the spheres to an existing interface from either bulk phase lowers the free energy, the creation of new interfaces to accommodate more spheres involves an increase of free energy. The spheres can considerably decrease the extra free energy which the emulsion must store, compared to the state in which the particles are present in the liquid phases, but they cannot gain for the emulsion a net free-energy advantage. However, once situated at the interface, the solid particles make coalescence of the droplets more difficult, essentially by keeping the droplets from coming into close contact. For two adjacent droplets, each carrying a close packed array of solid spheres to coalesce, some or all of the solid spheres have to escape into either of the bulk phases or, alternatively, the droplets will need to be sufficiently distorted to provide free surface for contact. Each of these processes requires an increase in free energy, providing a barrier to coalescence.

References

[1]   Tharwat Tadros, Applied Surfactants, Wiley-VCH, Germany, 2005

[2]   Tharwat Tadros, *Introduction to Surfactants*, de Gruyter, Germany, 2014.

[3]   W. C. Griffin, *J. Cosmet. Chemists* **1** (1949), 311; **5** (1954), 249.

[4]   J. T. Davies, *Proc. Int. Congr. Surface Activity* **1** (1959), 426.

[5]   J. T. Davies and E. K. Rideal, *Interfacial Phenomena*, Academic Press, New York, 1961.

[6]   H. Mollet and A. Grubenmann, *Formulation Technology*, Wiley-VCH, Germany, 2001.

[7]   Th. F. Tadros and B. Vincent, in: P. Becher (ed.), *Encyclopedia of Emulsion Technology*, Marcel Dekker, New York, 1983.

[8]   K. Shinoda, *J. Colloid Interface Sci.* **25** (1967), 396.

[9]   K. Shinoda and H. Saito, *J. Colloid Interface Sci.* **30** (1969), 258.

[10] K. Shinoda, *J. Chem. Soc. Japan* **89** (1968), 435.

[11] T. Tadros, *Rheology of Dispersions*, Wiley-VCH, Germany, 2010.

[12] A. Beerbower and M. W. Hill, *Amer. Cosmet. Perfum.* **87** (1972), 85.

[13] J. H. Hildebrand, *Solubility of Non-Electrolytes*, 2nd ed., Reinhold, New York, 1936.

[14] C. M. Hansen, *J. Paint Technol.* **39** (1967), 505.

[15] A. F. M. Barton, *Handbook of Solubility Prameters and Other Cohesive Parameters*, CRC Press, New York, 1983.

[16] J. N. Israelachvili, J. N. Mitchell and B. W. Ninham, *J. Chem. Soc., Faraday Trans. II*, **72** (1976), 1525.

[17] S. U. Pickering, *J. Chem. Soc.* (1934), 1112.

8 Creaming/sedimentation of emulsions and its prevention

8.1 Introduction

The process of creaming or sedimentation of emulsions is the result of gravity, when the density of the droplets and the medium are not equal. When the density of the disperse phase is lower than that of the medium (as with most oil-in-water O/W emulsions), creaming occurs, whereas if the density of the disperse phase is higher than that of the medium (as with most W/O emulsions), sedimentation occurs. Fig. 8.1 gives a schematic picture for creaming of emulsions for three cases [1–4].

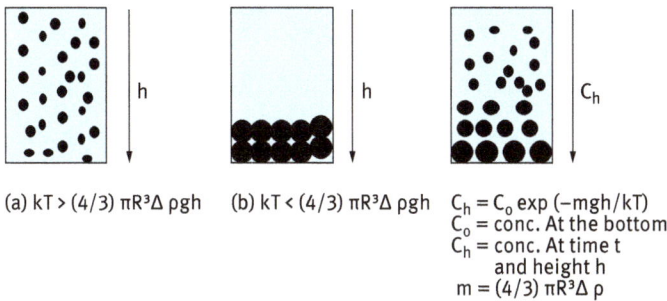

(a) $kT > (4/3) \pi R^3 \Delta \rho g h$ (b) $kT < (4/3) \pi R^3 \Delta \rho g h$ $C_h = C_o \exp(-mgh/kT)$
C_o = conc. At the bottom
C_h = conc. At time t
     and height h
$m = (4/3) \pi R^3 \Delta \rho$

Fig. 8.1: Representation of creaming of emulsions.

Case (a) represents the situation for small droplets ($< 0.1\,\mu m$, i.e. nano-emulsions) whereby the gravitational force is opposed by a diffusional force (i.e. associated with the translational kinetic energy of the droplets). A Boltzmann distribution is set up, whereby the droplet concentration C_h at height h is related to that at h = 0 by

$$C(h) = C_o \exp\left(-\frac{mgh}{kT}\right), \tag{8.1}$$

where mgh is the potential energy of a droplet at height h, with m being is the effective mass of a droplet, given by

$$m = \frac{4}{3}\pi R^3 \Delta \rho, \tag{8.2}$$

where R is the hydrodynamic radius of the droplets and $\Delta \rho$ is the density difference between the two liquid phases, k is the Boltzmann constant and T is the absolute temperature.

For no separation to occur, i.e. $C_h = C_o$,

$$kT \gg \frac{4}{3}\pi R^3 \Delta \rho g L. \tag{8.3}$$

Case (b) represents emulsions consisting of "monodisperse" droplets with radius $>1\,\mu m$. In this case, the concentration forces are much bigger than the opposing diffusional force, so that the emulsion separates into two distinct layers, with the droplets forming a cream or sediment leaving the clear supernatant liquid:

$$kT \ll \frac{4}{3}\pi R^3 \Delta \rho g L \,. \tag{8.4}$$

Case (c) is the case of a polydisperse (practical) emulsions, in which case the droplets will cream or sediment at various rates. In the last case, a concentration gradient builds-up, as predicted by equation (8.1), with the larger droplets staying at the top of the cream layer, with some smaller droplets remaining in the bottom layer.

8.2 Creaming or sedimentation rates

8.2.1 Very dilute emulsions ($\varphi < 0.01$)

In this case the rate could be calculated using Stokes' law which balances the hydro-dynamic force with gravity force [5]:

$$\text{Hydrodynamic Force} = 6\pi\eta_0 R v_0 \tag{8.5}$$

$$\text{Gravity Force} = \frac{4}{3}\pi R^3 \Delta \rho g \tag{8.6}$$

$$v_0 = \frac{2}{9}\frac{\Delta\rho g R^2}{\eta_0} \,; \tag{8.7}$$

v_0 is the Stokes' velocity and η_0 is the viscosity of the medium.

For an O/W emulsion with $\Delta\rho = 0.2$ in water ($\eta_0 \sim 10^{-3}$ Pa s), the rate of creaming or sedimentation is $\sim 4.4 \cdot 10^{-5}$ m s^{-1} for 10 μm droplets and $\sim 4.4 \cdot 10^{-7}$ m s^{-1} for 1 μm droplets. This means that in a 0.1 m container creaming or sedimentation of the 10 μm droplets is complete in ~ 0.6 h and for the 1 μm droplets this takes ~ 60 h.

If the droplets are deformable, a liquid drop moving within a second liquid phase has an internal circulation imparted to it. As a result of this, the motion through the continuous phase has a "rolling" as well as a "sliding" component. This situation has been treated theoretically resulting in the following equation [1]:

$$v = \frac{2}{3}\frac{\Delta\rho g R^2}{\eta_0}\frac{\eta_0 + \eta}{3\eta_0 + 2\eta} \,, \tag{8.8}$$

where η is the viscosity of the internal phase. For the case $\eta \gg \eta_0$, equation (8.8) predicts v to be 50 % higher than v_0, whereas for two liquids having similar viscosities ($\eta \sim \eta_0$), v is only 20 % greater than v_0. However, these theoretical predictions does not agree well with the experimental data for v. This could be due to the neglect of interfacial viscosity contribution.

With large droplets shape distortion may occur due to changes in pressure with "vertical height" of the droplet. The difference in pressure between the "top" and the "bottom" of the droplet is $\Delta\rho g d$, where d is the distorted vertical diameter. Any deviation from spherical geometry leads to an increase in surface area ΔA. Thus the distorting force due to gravity is opposed by the work necessary to increase the surface area $(\gamma\Delta A)$. For small deformities, $d \sim 2R$, and the fractional change in the radius is $\Delta\rho g R^2/\gamma$. For $\Delta\rho = 0.1\,\mathrm{g\,cm^3}$ and $\gamma = 2\,\mathrm{mN\,m^{-1}}$, a $2\,\mu m$ diameter droplet would undergo a distortion in radius of $\sim 5\cdot10^{-5}\,\%$, whereas for a $200\,\mu m$ diameter droplet the distortion would be $\sim 0.5\,\%$. Thus, this effect is really only significant for large droplets.

8.2.2 Moderately concentrated emulsions (0.2 > φ > 0.1)

As mentioned above, Stokes' law is only applicable for very dilute, non-interacting droplets. In most practical emulsions, the droplet concentration is high, and one has to take into account the hydrodynamic interaction between the droplets when $\varphi > 0.1$ (but less than 0.2). Three factors contributing to the change in creaming (or settling) rate when $\varphi > 0.1$ may be considered:

(i) The downward flux of fluid volume that accompanies the upward flux of oil droplets in order to maintain a zero mean volume flux at each point in a homogeneous emulsion. This change in fluid environment for one droplet causes the mean creaming rate to differ from its value at infinite dilution by an amount $-\varphi v_0$.

(ii) The second and largest contribution arises from the drag up of the fluid that adheres to the spherical droplets. This upward flux of fluid volume in the inaccessible shells surrounding the rigid spheres is accompanied by an equal downward flux in fluid volume in the part of the fluid that is accessible to the centre of a test sphere. In other words, the reduction in creaming rate arises from the diffuse downward current, which compensates for the upward flux in fluid volume in the inaccessible shell surrounding the rigid sphere. This contributes to $-4.5\varphi v_0$ to the change in the mean creaming velocity.

(iii) The third contribution to the change in creaming rate arises from the motion of the spheres, which generates collectively a velocity distribution in the fluid such that the second derivative of velocity, $\nabla^2 v$, has a nonzero mean. This property of the environment for a particular sphere changes its mean velocity by $0.5\varphi v_0$, i.e. it causes an increase in the creaming velocity.

(iv) The fourth contribution arises from the interaction between the spheres. When the test sphere whose velocity is being averaged is near one of the other spheres in the emulsion, the interaction between these two spheres gives the test sphere a velocity that is significantly different from that which is estimated from the velocity distribution in the absence of a second sphere. This gives a further change in the mean creaming rate of $-1.55\varphi v_0$. Therefore, the average velocity v can be related to that at infinite dilution v_0 (the Stokes' velocity) by taking into account the above four contributions:

$$v = v_0 + (-\varphi v_0 - 4.5\varphi v_0 + 0.5\varphi v_0 - 1.55\varphi v_0) = v_0(1 - 6.55\varphi). \qquad (8.9)$$

Equation (8.9) is referred to as the Bachelor's equation [6], which shows that for $\varphi = 0.1$, the rate of creaming or sedimentation is reduced by about 65 %.

8.2.3 Concentrated emulsions ($\varphi > 0.2$)

The rate of creaming or sedimentation becomes a complex function of φ, as is illustrated in Fig. 8.2, which also shows the change of relative viscosity η_r with φ. As can be seen from this figure, v decreases with an increase in φ, and ultimately it approaches zero when φ exceeds a critical value, φ_p, which is the so-called "maximum packing fraction". The value of φ_p for monodisperse "hard-spheres" ranges from 0.64 (for random packing) to 0.74 for hexagonal packing. The value of φ_p exceeds 0.74 for polydisperse systems. φ_p can also be much larger than 0.74 for emulsions which are deformable.

Fig. 8.2 also shows that when φ approaches φ_p, η_r approaches ∞. In practice most emulsions are prepared at φ values well below φ_p, usually in the range 0.2–0.5, and under these conditions creaming or sedimentation is the rule rather than the exception. Several procedures may be applied to reduce or eliminate creaming or sedimentation, and these are discussed below.

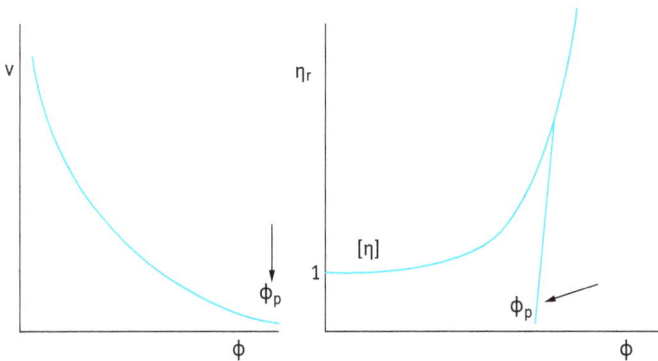

Fig. 8.2: Variation of v and η_r with φ.

8.3 Properties of a creamed layer

The structure of the creamed layer shown in Fig. 8.1 (c) represents the equilibrium volume, where the mutual distances between the droplets are determined by the balance between the external gravitational field and the mutual inter-droplet forces (electrostatic and/or steric forces associated with an interfacial polymer layer; see progressively lower. This shows that to produce physical gels from spherical geometry, the maximum packing in the creamed layer can exceed that for monodisperse spheres (φ = 0.74 for hexagonal packing and 0.64 for random packing). The effects of polydispersity are important in that the smaller droplets may fit into the voids between the larger droplets in a packed cream. Packings of greater than 0.90 can be achieved in this way. If the droplets are deformable, the volume of packed cream layer is reduced and φ values in the range 0.95–0.99 can be reached. The greater the size of the droplets and density difference between the two liquid phases, the greater the tendency for deformation to occur. In many cases, the droplets distort to polydedral cells resembling a foam in structure, with a corresponding network of more-or-less planer thin liquid films of one liquid separating cells of the other liquid. The stability of the system to coalescence then depends on the stability to rupture of these films, as will be discussed in Chapter 11.

If flocculation occurs in the emulsion (see Chapter 9) the overall rate of creaming will be faster, since the flocs are larger in size. However, the final cream volume will be greater due to the more open structure.

8.4 Prevention of creaming or sedimentation

8.4.1 Matching density of oil and aqueous phases

Clearly if $\Delta\rho$ = 0, v = 0, as shown by equation (8.7). Consider for example an O/W emulsion where the oil density is $0.9 \, \text{g cm}^{-3}$. To match the density of the aqueous phase ($\sim 1 \, \text{g cm}^{-3}$), the density of the oil must be increased; this can be achieved, for example, by replacing 20 % of the oil with another oil with density $1.4 \, \text{g cm}^{-3}$. Thus, this method is seldom practical. In addition if density matching is possible, it only occurs at one temperature.

8.4.2 Reduction of droplet size

Since the gravity force is proportional to R^3, then if R is reduced by a factor of 10, the gravity force is reduced by 1000. Below a certain droplet size (which also depends on the density difference between oil and water), the Brownian diffusion may exceed gravity, as shown by equation (8.3), and creaming or sedimentation is prevented. This is the principle of formulation of nanoemulsions (with size range 20–200 nm), which may show very little or no creaming or sedimentation.

8.4.3 Use of "thickeners"

These are high molecular weight polymers, natural or synthetic, such as Xanthan gum, hydroxyethyl cellulose, alginates, carragenans, etc. To understand the role of these "thickeners", let us consider the physical gels obtained by chain overlap. Flexible polymers that produce random coil in solution can produce "gels" at a critical concentration C^*, referred to as the polymer coil "overlap" concentration. This picture can be realised if one considers the coil dimensions in solution. Considering the polymer chain to be represented by a random walk in three dimensions, one may define two main parameters, namely the root mean square end-to-end length $\langle r^2 \rangle^{1/2}$ and the root mean square radius of gyration $\langle s^2 \rangle^{1/2}$ (sometimes denoted by R_G). The two are related by

$$\langle r^2 \rangle^{1/2} = 6^{1/2} \langle s^2 \rangle^{1/2} \tag{8.10}$$

The viscosity of a polymer solution increases gradually with increase in its concentration and at a critical concentration, C^*, the polymer coils with a radius of gyration R_G and a hydrodynamic radius R_h (which is higher than R_G due to solvation of the polymer chains) begin to overlap, and this shows a rapid increase in viscosity. This is illustrated in Fig. 8.3, which shows the variation of $\log \eta$ with $\log C$.

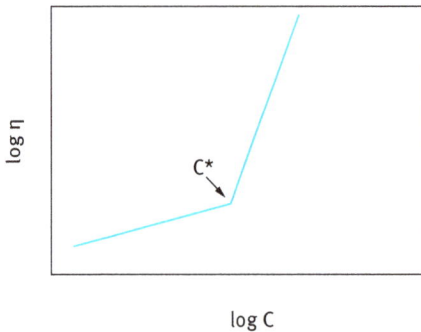

log C   Fig. 8.3: Variation of $\log \eta$ with $\log C$.

In the first part of the curve $\eta \propto C$, whereas in the second part (above C^*) $\eta \propto C^{3.4}$. A schematic representation of polymer coil overlap is shown in Fig. 8.4, which shows the effect of gradually increasing the polymer concentration. The polymer concentration above C^* is referred to as the semi-dilute range.

C* is related to R_G and the polymer molecular weight M by

$$C^* = \frac{3M}{4\pi R_G^3 N_{av}}, \tag{8.11}$$

where N_{av} is the Avogadro number. As M increases C^* becomes progressively lower. This shows that to produce physical gels at low concentrations by simple polymer coil overlap one has to use high molecular weight polymers. Another method to reduce the polymer concentration at which chain overlap occurs is to use polymers that form extended chains such as xanthan gum, which produces conformation in the form of a helical structure with a large axial ratio. These polymers give much higher intrinsic viscosities, and they show both rotational and translational diffusion. The relaxation time for the polymer chain is much higher than a corresponding polymer with the same molecular weight but produces random coil conformation.

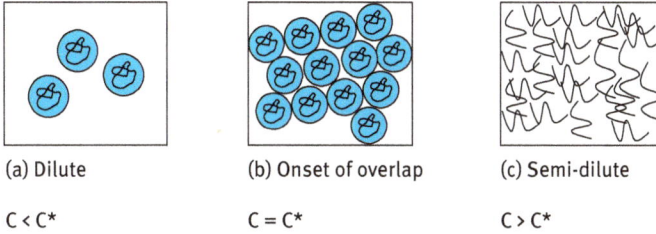

(a) Dilute   (b) Onset of overlap   (c) Semi-dilute

$C < C^*$ $C = C^*$ $C > C^*$

Fig. 8.4: Cross-over between dilute and semi-dilute solutions.

The above polymers interact at very low concentrations, and the overlap concentration can be very low ($< 0.01\,\%$). These polysaccharides are used in many formulations to produce physical gels at very low concentrations. At sufficiently high polymer concentration, the elastic properties of the physical gels can arrest the oil droplets, thus preventing creaming or sedimentation.

Let us consider the gravitational stresses exerted during creaming or sedimentation:

$$\text{Stress} = \text{mass of drop} \cdot \text{acceleration of gravity} = \frac{4}{3}\pi R^3 \Delta\rho g. \tag{8.12}$$

To overcome such stress one needs a restoring force:

$$\text{Restoring force} = \text{area of drop} \cdot \text{stress of drop} = 4\pi R^2 \sigma_p. \tag{8.13}$$

Thus, the stress exerted by the droplet σ_p is given by

$$\sigma_p = \frac{\Delta\rho R g}{3}. \tag{8.14}$$

The maximum shear stress developed by an isolated spherical droplet through a medium of viscosity η is given by the expression [5]

$$\sigma_p = \frac{3v\eta}{2R}. \tag{8.15}$$

For droplets at the coarse end of the colloidal range, σ_p is in the range 10^{-5}–10^{-2} Pa. As an illustration Fig. 8.5 shows the variation of viscosity with shear stress for a typical thickener, namely ethylhydroxethyl cellulose (EHEC).

The results of Fig. 8.6 show that below a critical value of the shear stress ($\sim 0.2\,\text{Pa}$) the viscous behaviour is Newtonian. Above this stress value the viscosity decreases,

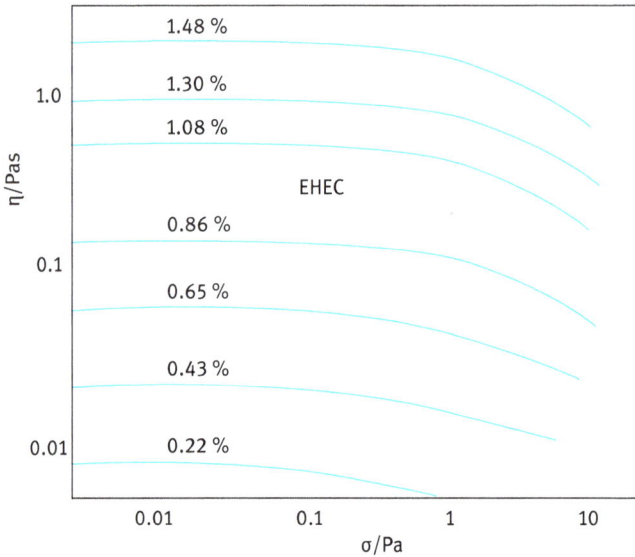

Fig. 8.5: Viscosity as a function of shear stress for various EHEC concentrations.

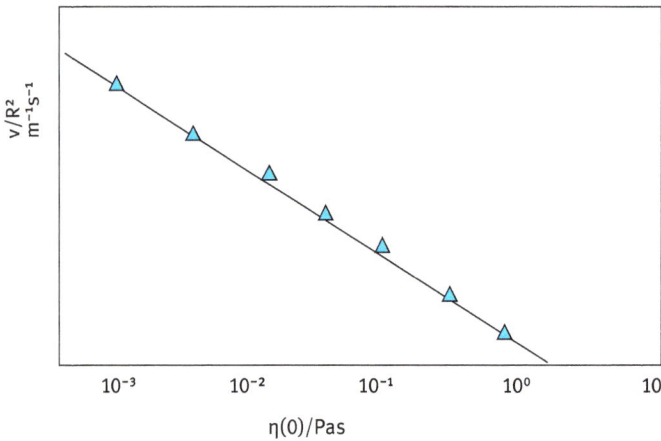

Fig. 8.6: Sedimentation rate vs $\eta(0)$.

indicating the shear thinning behaviour. The plateau value at $\sigma < 0.2$ gives the limiting and residual viscosity $\eta(0)$, sometimes referred to as the zero shear viscosity. Thus for systems containing thickeners, the viscosity of the medium in equation (8.7) may be replaced by the residual viscosity $\eta(0)$. This results in considerable reduction of the creaming or sedimentation rate. For example for a very dilute emulsion containing 1.48 % EHEC with $\eta(0) \sim 2$ Pa s, the rate of creaming for 10 µm droplets with $\Delta\rho = 0.2$ g cm^{-3} is $\sim 2.2 \cdot 10^{-8}$ m s^{-1}, which is about three orders of magnitude lower than the value in water ($\sim 4.4 \cdot 10^{-5}$ m s^{-1}). This means that in a 10 cm tube complete creaming will take more than a month. For a 1 µm droplet, complete creaming time will take about two years. In more concentrated emulsions this creaming time will be much longer.

The above analysis implies that the residual viscosity $\eta(0)$ can be used to predict creaming or sedimentation of emulsions. This is illustrated in Fig. 8.6, which shows a plot of v/R^2 (m^{-1} s^{-1}) versus $\eta(0)$ for a latex with volume fraction 0.05 in solutions of EHEC. A linear correlation between v/R^2 and $\eta(0)$ is clearly shown.

8.4.4 Reduction of creaming/sedimentation of emulsions using associative thickeners

Associative thickeners are hydrophobically modified polymer molecules, whereby alkyl chains (C_{12}–C_{16}) are either randomly grafted onto a hydrophilic polymer molecule such as hydroxyethyl cellulose (HEC) or simply grafted at both ends of the hydrophilic chain. An example of hydrophobically modified HEC is Natrosol plus (Hercules) which contains 3–4 C_{16} randomly grafted onto hydroxyethyl cellulose. Another example of a polymer that contains two alkyl chains at both ends of the molecule is HEUR (Rohm and Haas), which is made of polyethylene oxide (PEO) capped at both ends with a linear C_{18} hydrocarbon chain.

The above hydrophobically modified polymers form gels when dissolved in water. Gel formation can occur at relatively lower polymer concentrations compared with the unmodified molecule. The most likely explanation of gel formation is due to hydrophobic bonding (association) between the alkyl chains in the molecule. This effectively causes an apparent increase in the molecular weight. These associative structures are similar to micelles, except the aggregation numbers are much smaller.

Fig. 8.7 shows the variation of viscosity (measured using a Brookfield at 30 rpm as a function of the alkyl content (C_8, C_{12} and C_{16}) for hydropphobically modified HEC (i.e HMHEC). The viscosity reaches a maximum at a given alkyl group content that decreases with increase in the alkyl chain length. The viscosity maximum increases with increase in the alkyl chain length.

Associative thickeners also show interaction with surfactant micelles that are present in the formulation. The viscosity of the associative thickeners shows a maximum at a given surfactant concentration which depends on the nature of surfactant. This is shown schematically in Fig. 8.8. The increase in viscosity is attributed to the hydrophobic interaction between the alkyl chains on the backbone of the polymer with the surfactant micelles. A schematic picture showing the interaction between HM polymers and surfactant micelles is shown in Fig. 8.9. At higher surfactant concentration, the "bridges" between the HM polymer molecules and the micelles are broken (free micelles) and η decreases.

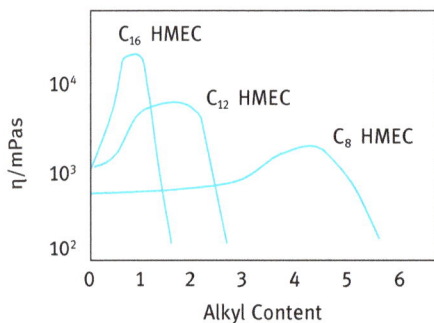

Fig. 8.7: Variation of viscosity of 1 % HMHEC vs alkyl group content of the polymer.

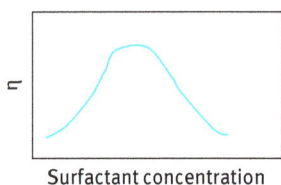

Fig. 8.8: Schematic plot of viscosity of HM polymer with surfactant concentration.

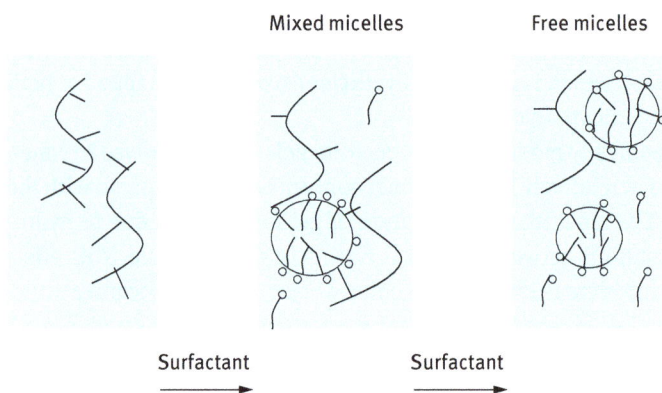

Fig. 8.9: Schematic representation of the interaction of polymers with surfactants.

The viscosity of hydrophobically modified polymers shows a rapid increase a critical concentration, which may be defined as the critical aggregation concentration (CAC) as is illustrated in Fig. 8.10 for HMHEC (WSP-D45 from Hercules). The assumption is made that the CAC is equal to the coil overlap concentration C^*.

From a knowledge of C^* and the intrinsic viscosity $[\eta]$ one can obtain the number of chains in each aggregate. For the above example $[\eta] = 4.7$ and $C^*[\eta] = 1$ giving an aggregation number of ~ 4

At C^* the polymer solution shows non-Newtonian flow (shear thinning behaviour) and a high viscosity at low shear rates. This is illustrated in Fig. 8.11, which shows the variation of apparent viscosity with shear rate (using a constant stress rheometer).

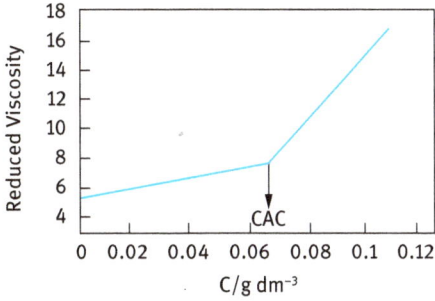

Fig. 8.10: Variation of reduced viscosity with HMHEC concentration.

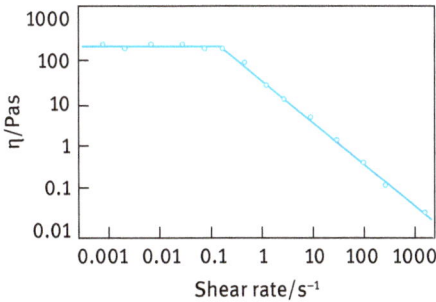

Fig. 8.11: Variation of viscosity with shear rate for HMEC WSP-47 at 0.75 g/100 cm³.

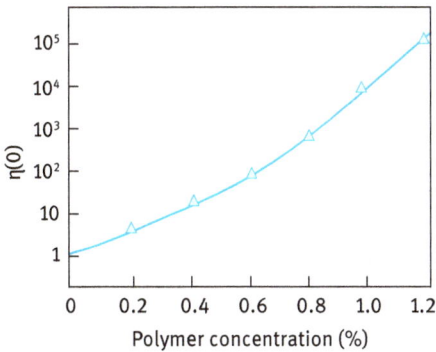

Fig. 8.12: Variation of $\eta(0)$ with polymer concentration.

Below $\sim 0.1\,\mathrm{s}^{-1}$, a plateau viscosity value $\eta(0)$ (referred to as residual or zero shear viscosity) is reached ($\sim 200\,\mathrm{Pa\,s}$).

With an increase in polymer concentrations above C* the zero shear viscosity increases with increase in polymer concentration. This is illustrated in Fig. 8.12

The above hydrophobically modified polymers are viscoelastic; this is illustrated in Fig. 8.13 for a solution 5.25 % of C_{18} end-capped PEO with M $=$ 35 000, which shows the variation of the storage modulus G′ and loss modulus G″ with frequency ω ($\mathrm{rad\,s}^{-1}$). G′ increases with increase in frequency and ultimately reaches a plateau value at high frequency. G″ (which is higher than G′ in the low frequency regime) increases with increase in frequency, reaches a maximum at a characteristic fre-

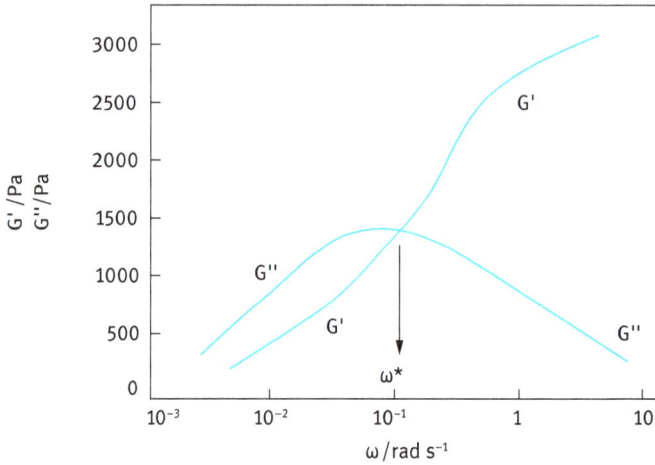

Fig. 8.13: Variation of G′ and G″ with frequency for 5.24 HM PEO.

quency ω^* (at which G′ = G″) and then decreases to near zero value in the high frequency regime.

The above variation of G' and G" with ω is typical for a system that shows Maxwell behaviour.

From the cross-over point ω^* (at which G′ = G″) one can obtain the relaxation time τ of the polymer in solution:

$$\tau = \frac{1}{\omega^*}.\tag{8.16}$$

For the above polymer τ = 8 s.

The above gels (sometimes referred to as rheology modifiers) are used for reduction of creaming or sedimentation of emulsions. These hydrophobically modified polymers can also interact with hydrophobic oil droplets in an emulsion forming several other associative structures.

8.4.5 Controlled flocculation

As discussed before, the total energy-distance of separation curve for electrostatically stabilised shows a shallow minimum (secondary minimum) at relatively long distance of separation between the droplets. By addition of small amounts of electrolyte, such a minimum can be made sufficiently deep for weak flocculation to occur. The same applies to sterically stabilised emulsions, which show only one minimum, whose depth can be controlled by reducing the thickness of the adsorbed layer. This can be obtained by reducing the molecular weight of the stabiliser and/or addition of a non-solvent for the chains (e.g. electrolyte).

This phenomenon of weak flocculation may be applied to reduce creaming or sedimentation, in particular for concentrated emulsions. In the latter case, the attractive energy required for inducing weak flocculation can be small (of the order of few kT units). This can be understood if one considers the free energy of flocculation, which consists of two terms: an energy term determined by the depth of the minimum (G_{min}) and an entropy term that is determined by reduction in configurational entropy on aggregation of droplets:

$$\Delta G_{flocc} = \Delta H_{flocc} - T\Delta S_{flocc} .\tag{8.17}$$

With concentrated emulsions, the entropy loss on flocculation is small compared with that for a dilute emulsion. Hence, for flocculation of a concentrated emulsion a small energy minimum is sufficient compared with the case with a dilute emulsion.

8.4.6 Depletion flocculation

As mentioned above, many thickeners such as HEC or xanthan gum may not adsorb on the droplets. This is described as "free" (non-adsorbing) polymer in the continuous phase [9]. At a critical concentration, or volume fraction of free polymer, φ_p^+, weak flocculation occurs, since the free polymer coils become "squeezed-out" from between the droplets. This is illustrated in Fig. 8.15, which sows the situation when the polymer volume fraction exceeds the critical concentration. Since the polymer coils do not adsorb on the droplet surface, a "polymer-free" zone with thickness Δ (proportional to the radius of gyration R_G of the free polymer) is produced. When two droplets approach each other such that the interdroplet distance $h \leq 2\Delta$, the polymer coils become "squeezed out" from between the droplets, as illustrated in Fig. 8.14.

The osmotic pressure outside the droplets is higher than in between the droplets, and this results in attraction the magnitude of which depends on the concentration of the free polymer and its molecular weight, as well as the droplet size and φ. The value of φ_p^+ decreases with an increase of the molecular weight of the free polymer. It also decreases as the volume fraction of the emulsion increases.

Initial addition of "free polymer" produces weak flocs, which shows an increase in the rate of creaming. With further increase in the "free polymer" concentration the rate of creaming increases, and the emulsion separates into two layers: a cream layer at the top and a clear liquid layer at the bottom of the container. However, above a critical concentration of "free polymer", the emulsion shows no creaming. This behaviour was recently demonstrated by Abend et al. [10], who investigated the effect of the addition of xanthan gum to an O/W emulsion (the oil being Isopar V, a hydrocarbon oil) with a volume fraction $\varphi = 0.5$, which was stabilised using an A–B–A block copolymer (with A being polyethylene oxide, PEO, and B being polypropylene oxide, PPO, containing 26.5 EO units and 29.7 PO units). The xanthan gum concentration var-

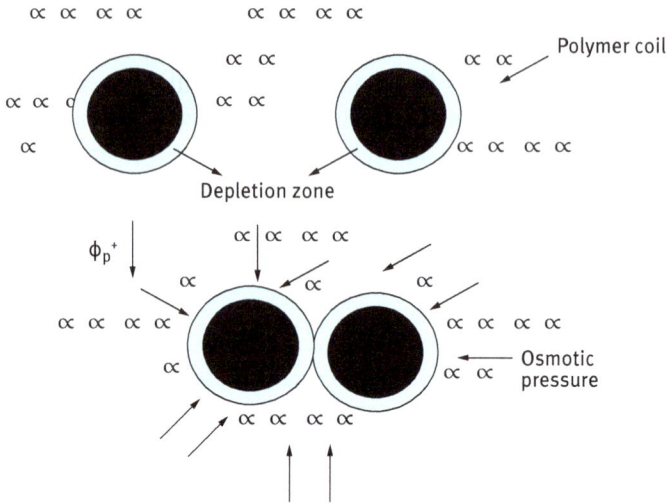

Fig. 8.14: Schematic representation of depletion flocculation.

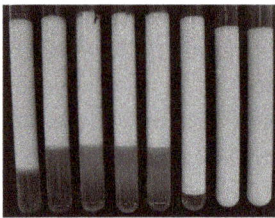

Fig. 8.15: Visual observation (after 8 months storage) of O/W emulsions ($\varphi = 0.5$) stabilised with an A–B–A block copolymer (PEO–PPO–PEO) at various xanthan gum concentrations; from left to right: 0.0, 0.01, 0.05, 0.1, 0.25, 0.50, 0.67 % xanthan gum.

ied between 0.01 and 0.67 %. Fig. 8.15 shows the visual observation of the emulsions after 8 months of storage.

It can be seen from Fig. 8.15 that an initial addition of xanthan gum up to 0.1 % enhances the creaming of the emulsion as a result of depletion flocculation. The emulsion containing 0.25 % xanthan gum shows less creaming, but phase separation did occur. However at 0.5 and 0.67 % xanthan gum there is no creaming after storage for 8 months. At a concentration of 0.5 and 0.67 % xanthan gum, the arrested network formed by the droplets appears now to be strong enough to withstand the gravitational stress (creaming), and no clear water-phase is visible at the bottom of the vessel even after eight months of storage.

Thus, the above weak flocculation can be applied to reduce creaming or sedimentation although it suffers from the following drawbacks: (i) temperature dependence; as the temperature increases, the hydrodynamic radius of the free polymer decreases (due to dehydration), and hence more polymer will be required to achieve the same effect at lower temperatures; (ii) if the free polymer concentration is below a certain limit, phase separation may occur, and the flocculated emulsion droplets may cream or sediment faster than in the absence of the free polymer.

8.4.7 Use of "inert" fine particles

Several fine particulate inorganic material produce "gels" when dispersed in aqueous media, e.g. sodium montmorillonite or silica. These particulate materials produce three dimensional structures in the continuous phase as a result of interparticle interaction. For example, sodium montmorillonite (referred to as swellable clays) form gels at low and intermediate electrolyte concentrations. This can be understood from a knowledge of the structure of the clay particles. The latter consist of plate-like particles consisting of a octahedral alumina sheet sandwiched between two tetrahedral silica sheets. In the tetrahedral sheet tetravalent Si is sometimes replaced by trivalent Al. In the octahedral sheet there may be replacement of trivalent Al by divalent Mg, Fe, Cr or Zn. The small size of these atoms allows them to take the place of small Si and Al. This replacement is usually referred to as isomorphic substitution, whereby an atom of lower positive valence replaces one of higher valence, resulting in a deficit of positive charge or excess of negative charge. This excess of negative layer charge is compensated by adsorption at the layer surfaces of cations that are too big to be accommodated in the crystal. In aqueous solution, the compensation cations on the layer surfaces may be exchanged by other cations in solution, and hence may be referred to as exchangeable cations. With montmorillonite, the exchangeable cations are located on each side of the layer in the stack, i.e. they are present in the external surfaces as well as between the layers. When montrmorillonite clays are placed in contact with water or water vapour, the water molecules penetrate between the layers, causing interlayer swelling or (intra)crystalline swelling. This interlayer swelling leads, at most, to doubling of the volume of dry clay where four layers of water are adsorbed. The much larger degree of swelling, which is the driving force for "gel" formation (at low electrolyte concentration) is due to osmotic swelling. It has been suggested that swelling of montmorillonite clays is due to the electrostatic double layers that are produced between the charge layers and cations. This is certainly the case at low electrolyte concentration where the double layer extension (thickness) is large.

As discussed above, the clay particles carry a negative charge as a result of isomorphic substitution of certain electropositive elements by elements of lower valency. The negative charge is compensated by cations, which in aqueous solution form a diffuse layer, i.e. an electric double layer is formed at the clay plate/solution interface. This double layer has a constant charge, which is determined by the type and degree of isomorphic substitution. However, the flat surfaces are not the only surfaces of the plate-like clay particles – they also expose an edge surface. The atomic structure of the edge surfaces is entirely different from that of the flat-layer surfaces. At the edges, the tetrahedral silica sheets and the octahedral alumina sheets are disrupted, and the primary bonds are broken. The situation is analogous to that of the surface of silica and alumina particles in aqueous solution. On such edges, therefore, an electric double layer is created by adsorption of potential determining ions (H^+ and OH^-), and one may, therefore identify an isoelectric point (i.e.p.) as the point of zero charge (p.z.c.) for these edges. With broken octahedral sheets at the edge, the surface behaves as Al–OH with an i.e.p in the region of pH 7–9. Thus in most cases the edges become negatively charged above pH 9 and positively charged below pH 9.

Van Olphen [11] suggested a mechanism of gel formation of montmorillonite involving interaction of the oppositely charged double layers at the faces and edges of the clay particles. This structure, which is usually referred to as a "card-house" structure, was considered to be the reason for the formation of the voluminous clay gel. However, Norrish [12] suggested that the voluminous gel is the result of the extended double layers, particularly at low electrolyte concentrations. A schematic picture of gel formation produced by double layer expansion and "card-house" structure is shown in Fig. 8.16.

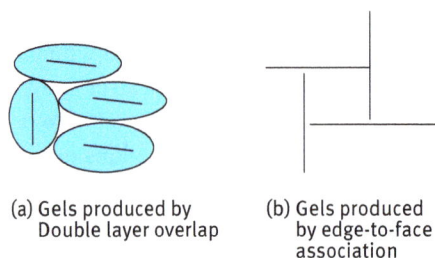

(a) Gels produced by Double layer overlap

(b) Gels produced by edge-to-face association

Fig. 8.16: Schematic representation of gel formation in aqueous clay dispersions.

Finely divided silica such as Aerosil 200 (produced by Degussa) produce gel structures by simple association (by van der Waals attraction) of the particles into chains and cross-chains. When incorporated in the continuous phase of an emulsion, these gels prevent creaming or sedimentation.

8.4.8 Use of mixtures of polymers and finely divided particulate solids

By combining the thickeners such as hydroxyethyl cellulose or xanthan gum with particulate solids such as sodium montmorrillonite, a more robust gel structure could be produced. By using such mixtures, the concentration of the polymer can be reduced, thus overcoming the problem of dispersion on dilution. This gel structure may be less temperature dependent and could be optimised by controlling the ratio of the polymer and the particles. If these combinations of say sodium montmorillonite and a polymer such as hydroxyethly cellulose, polyvinyl alcohol (PVA) or xanthan gum, are balanced properly, they can provide a "three-dimensional structure" which entraps all the droplets and stop creaming or sedimentation of the emulsion. The mechanism of gelation of such combined systems depends to a large extent on the nature of the droplets, the polymer and the conditions. If the polymer adsorbs on the particle surface (e.g. PVA on sodium montmorillonite or silica) a three-dimensional network may be formed by polymer bridging. Under conditions of incomplete coverage of the particles by the polymer, the latter becomes simultaneously adsorbed on two or more particles. In other words the polymer chains act as "bridges" or "links" between the particles.

8.4.9 Use of liquid crystalline phases

Surfactants produce liquid crystalline phases at high concentrations [3]. Three main types of liquid crystals can be identified, as illustrated in Fig. 8.17: hexagonal phase (sometimes referred to as middle phase), cubic phase and lamellar (neat phase). All these structures are highly viscous and they also show elastic response. If produced in the continuous phase of emulsions, they can eliminate creaming or sedimentation of the droplets. These liquid crystalline phase are particularly useful for application in hand creams which contain high surfactant concentrations.

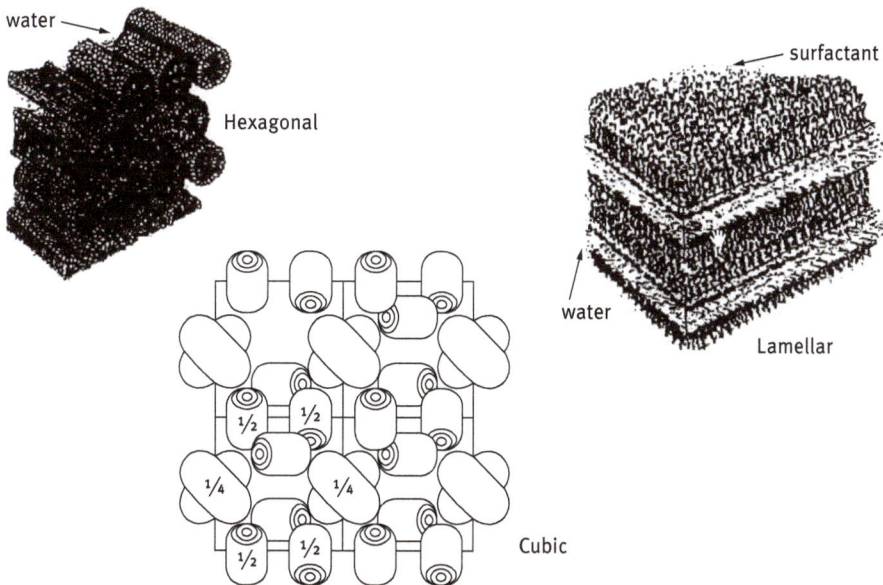

Fig. 8.17: Schematic picture of liquid crystalline phases.

References

[1]   Th. F. Tadros and B. Vincent, in: P. Becher (ed.), *Encyclopedia of Emulsion Technology*, Marcel Dekker, New York, 1983.
[2]   B. P. Binks (ed.), *Modern Aspects of Emulsion Science*, The Royal Society of Chemistry Publication Cambridge, 1998.
[3]   Tharwat Tadros, *Applied Surfactants*, Wiley-VCH, Germany, 2005.
[4]   Tharwat Tadros (ed.), *Emulsion Formation and Stability*, Wiley-VCH, Germany, 2013.
[5]   Th. F. Tadros, in: Th. F. Tadros (ed.), *Solid/Liquid Disperseions*, Academic Press, London, 1967.
[6]   G. K. Batchelor, *J. Fluid. Mech.* **52** (1972), 245.
[7]   R. Buscall, J. W. Goodwin, R. H. Ottewill and Th. F. Tadros, *J. Colloid Interface Sci.* **85** (1982), 78.
[8]   Tharwat Tadros, *Rheology of Dispersions*, Wiley-VCH, Germany, 2010.
[9]   S. Asakura and F. Oosawa, *J. Phys. Chem.* **22** (1954), 1255; *J. Polym. Sci.* **33** (1958), 183.
[10] S. Abend, C. Holtze, T. Tadros and P. Schutenberger, *Langmuir* **28** (2012), 7967.
[11] H. van Olphen, *Clay Colloid Chemistry*, Wiley, New York, 1963.
[12] K. Norrish, *Discussion Faraday Soc.* **18** (1954) , 120.

9 Flocculation of emulsions

9.1 Introduction

Flocculation is the process in which the emulsion drops aggregate, without rupture of the stabilising layer at the interface, if the pair interaction free energy becomes appreciably negative at a certain separation. This negative interaction is the result of van der Waals attraction G_A that is universal for all disperse systems. As shown in Chapter 3, G_A is inversely proportional to the droplet-droplet distance of separation h, and it depends on the effective Hamaker constant A of the emulsion system. Flocculation may be weak (reversible) or strong (not easily reversible), depending on the strength of the inter-droplet forces. Flocculation usually leads to enhanced creaming, because the flocs rise faster than individual drops, due to their larger effective radius. Exceptions occur in concentrated emulsions, where the formation of gel-like network structure can have a stabilising influence (see Chapter 8). Flocculation is enhanced by polydispersity since the differential creaming speeds of small and large drops cause them to come to close proximity more often than would occur in a monodisperse system [1]. The cream layer formed towards the end of the creaming process is actually a concentrated floc. The rate of flocculation can be estimated from the product of a frequency factor (how often drops encounter each other) and a probability factor (how long they stay together). The former can be calculated for the case of Brownian motion (perikinetic flocculation) or under shear flow (orhokinetic flocculation), as will be discussed below. The orthokinetic flocculation depends on the interaction energy, i.e. the free energy required to bring drops from infinity to a specified distance apart.

In calculating the interaction energy as a function of inter-droplet distance, three terms are normally considered (see Chapter 3): (i) van der Waals attraction (which depends on droplet radius and the effective Hamaker constant), (ii) electrostatic repulsion, produced for example by adsorption of ionic surfactants (which depends on the surface or zeta potential, drop radius and ionic strength of the medium) and (iii) steric repulsion, produced by adsorption of non-ionic surfactants or polymers (which depends on the adsorption density, conformation of the chain at the O/W interface and solvent quality).

9.2 Mechanism of emulsion flocculation

This can occur if the energy barrier is small or absent (for electrostatically stabilised emulsions) or when the stabilising chains reach poor solvency (for sterically stabilised emulsions, i.e. the Flory–Huggins interaction parameter $\chi > 0.5$). For convenience, I will discuss flocculation of electrostatically and sterically stabilised emulsions separately.

9.2.1 Flocculation of electrostatically stabilised emulsions

As discussed in Chapter 3, the condition for kinetic stability described by the Deyaguin–Landua–Verwey–Overbeek (DLVO) theory [2, 3] is the magnitude of the energy maximum, G_{max}, at intermediate separation of droplets. For an emulsion to remain kinetically stable (with no flocculation), $G_{max} > 25$ kT. When $G_{max} < 5$ kT, or completely absent, flocculation occurs. Two types of flocculation kinetics may be distinguished: fast flocculation with no energy barrier, and slow flocculation when an energy barrier exists.

Fast flocculation kinetics

The fast flocculation kinetics was treated by Smoluchowki [4], who considered the case where there is no interaction between the two colliding droplets until they come into contact, whereupon they adhere irreversibly. The process is simply represented by second-order kinetics and is simply diffusion controlled. The number of particles n at any time t may be related to the finial number (at t = 0) no by the following expression:

$$n = \frac{n_0}{1 + k_0 n_0 t} \, , \tag{9.1}$$

where k_0 is the rate constant for fast flocculation related to the diffusion coefficient of the droplets D, i.e.

$$k_0 = 8 \pi D R \, . \tag{9.2}$$

D is given by the Stokes–Einstein equation:

$$D = \frac{kT}{6 \pi \eta R} \, . \tag{9.3}$$

Combining equations (9.2) and (9.3),

$$k_0 = \frac{4}{3} \frac{kT}{\eta} = 5.5 \cdot 10^{-18} \, \mathrm{m^3 \, s^{-1}} \text{ for water at } 25\,^\circ\mathrm{C}. \tag{9.4}$$

The half-life $t_{1/2}$ ($n = \frac{1}{2} n_0$) can be calculated at various n_0 or volume fraction φ as given in Tab. 9.1.

Tab. 9.1: Half-life of emulsion flocculation.

R / µm	φ			
	10^{-5}	10^{-2}	10^{-1}	$5 \cdot 10^{-1}$
0.1	765 s	76 ms	7.6 ms	1.5 ms
1.0	21 h	76 s	7.6 s	1.5 s
10.0	4 months	21 h	2 h	25 m

It can be seen from Tab. 9.1 that the time scales involved at various droplet sizes and volume factions range over some 10 orders of magnitude. A dilute emulsion of large droplets may show no visible sign of flocculation over a day or so, whereas a high concentration of small droplets would appear to be instantly flocculated.

The Smoluchowski analysis also leads to an expression for the number of -mer aggregates at time t:

$$n_i = \frac{n_o(t/t_{1/2})^{i-1}}{(1 + t/t_{1/2})^{i+1}} . \qquad (9.5)$$

Most emulsions are polydisperse, and this causes an increase in flocculation rate compared to that of monodisperse emulsions. Another factor that affects the rate of fast flocculation arises from hydrodynamic effects, which show a dependence on droplet size and number concentration. The maximum value of k_o found for these systems is $3 \cdot 10^{-18}$ m^3 s^{-1}, compared to the Smoluchowski value of $5.5 \cdot 10^{-18}$ m^3 s^{-1}. This is due to the fact that the hydrodynamic interactions for diffusing droplets obeys Stokes' law only when the droplets are isolated from each other. In a real emulsion, extra hydrodynamic interactions have to be taken into account. Thus, for approaching droplets, the assumption made in the Smoluchowski analysis that the relative diffusion coefficient D_{12} is given by the sum of their two individual diffusion coefficients D_1 and D_2 is no longer valid.

Slow flocculation kinetics

Smoluchowski [4] originally accounted for the effect of an energy barrier G_{max}, arising from interparticle interaction, on the kinetics of flocculation by introducing a correction parameter α, where α is the fraction of collisions which are "effective", i.e. leading to irreversible flocculation. This idea was based on the Arrhenius equation for chemical reactions, i.e.

$$k = k_o \exp\left(-\frac{G_{max}}{kT}\right). \qquad (9.6)$$

The analogy with chemical reactions is not exact. In chemical kinetics the rate of disappearance of a reactant is given by the absolute concentration of the transition-state species times the frequency of the bond in that species which has to break to form the products. In flocculation kinetics the rate depends on the flux of particles to any chosen particle.

Fuchs [5] showed when particle interactions are present the flux is made up of two contributions, one due to the Brownian diffusion of the particles, the other due to the interactions. In this way, the rate constant k of slow flocculation is related to the Smoluchowski rate k_o by the stability constant W:

$$W = \frac{k_o}{k} . \qquad (9.7)$$

W is related to G_{max} by the following expression [6]:

$$W = 2R \int_{2R}^{\infty} \exp\left(\frac{G_{max}}{kT}\right) r^{-2} \, dr, \qquad (9.8)$$

where r equals the centre–centre separation of two interacting droplets. W depends on the extent of flocculation and the morphology of the flocs in a given system. For this reason it is usual only to consider the early stages of flocculation ($t \to 0$), where only doublets are the only aggregated structures.

For charge-stabilised emulsions, W is given by the following expression [6]:

$$W = \frac{1}{2\kappa R} \exp\left(\frac{G_{max}}{kT}\right), \qquad (9.9)$$

where κ is the Debye–Huckel parameter, which is given by

$$\kappa = \left(\frac{2Z^2 e^2 C}{\varepsilon_r \varepsilon_0 kT}\right)^{1/2}, \qquad (9.10)$$

where Z is the valency of counter-ions, e is the electronic charge, C is the electrolyte concentration in bulk solution, ε_r is the relative permittivity of the medium, ε_0 is the permittivity of free space, k is the Boltzmann constant and T is the absolute temperature.

Since G_{max} is determined by the electrolyte concentration C and valency, one can derive an expression relating W to C and Z:

$$\log W = \text{const.} - 2.06 \cdot 10^9 \left(\frac{R\gamma^2}{Z^2}\right) \log C, \qquad (9.11)$$

where γ is a function that is determined by the surface potential ψ_0:

$$\gamma = \left[\frac{\exp(Ze\psi_0/kT) - 1}{\exp(ZE\psi_0/kT) + 1}\right]. \qquad (9.12)$$

Plots of $\log W$ vs $\log C$ are shown in Fig. 9.1, which shows a linear relationship in the slow flocculation regime. In the fast flocculation regime $G_{max} = 0$ and $d(\log W)/$

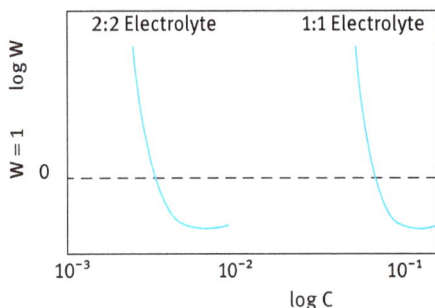

Fig. 9.1: log W–log C curves for electrostatically stabilised emulsions.

$d(\log C) = 0$. The condition $\log W = 0$ ($W = 1$) is the onset of fast flocculation. The electrolyte concentration at this point defines the critical flocculation concentration CFC. Above the CFC $W < 1$ (due to the contribution of van der Waals attraction which accelerates the rate above the Smoluchowski value). Below the CFC, $W > 1$ and it increases with decrease of electrolyte concentration. The figure also shows that the CFC decreases with increase of valency in accordance to the Schulze–Hardy rule.

Weak (reversible) flocculation

As described in Chapter 3, the energy–distance curve described by the DLVO theory [2, 3] shows the presence of a shallow energy well, namely the secondary minimum (G_{min}), which is few kT units. In this case flocculation is weak and reversible, and hence one must consider both the rate of flocculation (forward rate k_f) and defloccu-lation (backward rate k_b). The rate of decrease of particle number with time is given by the expression

$$-\frac{dn}{dt} = -k_f n^2 + k_b n.$$ (9.13)

The backward reaction (break-up of weak flocs) reduces the overall rate of floccula-tion. k_b may depend on the floc size and the exact way in which the flocs break down. Another complication in the analysis of weak (reversible) flocculation is the effect of droplet number concentration. Flocculation of this type is a critical phenomenon rather than a chain (or sequential) process. Thus, a critical droplet number concen-tration, n_{crit} has to be exceeded before flocculation occurs.

Orthokinetic flocculation

This process of flocculation occurs under shearing conditions and is referred to as orthokinetic (to distinguish it from the diffusion controlled perikinetic process). The simplest analysis is for laminar flow, since for turbulent flow with chaotic vortices (as is the case in a high-speed mixer) the particles are subjected to a wide and unpre-dictable range of hydrodynamic forces [7]. For laminar flow, the particle will move at the velocity of the liquid at the plane coincident with the centre of the particle, v_p. In this case the total collision frequency due to flow, c_f, is given by the following expres-sion:

$$c_f = \frac{16}{3} n_p^2 R^3 \left(\frac{dv}{dx} \right).$$ (9.14)

As the particles approach the shear field, the hydrodynamic interactions cause the colliding pair to rotate, and with the combination of the slowing approach due to liq-uid drainage (lubrication stress) and Brownian motion, not all collisions will lead to aggregation. Equation (9.14) must be reduced by a factor α (the collision frequency) to account for this:

$$c_f = \alpha \frac{16}{3} n_p^2 R^3 \left(\frac{dv}{dx} \right).$$ (9.15)

The collision frequency α is of the order 1, and a typical value would be $\alpha \approx 0.8$.

(dv/dx) is the shear rate so that equation (9.15) can be written as

$$c_f = \alpha \frac{16}{3} n_p^2 R^3 \dot{\gamma},$$ (9.16)

and the rate of orthokinetic flocculation is given by

$$-\frac{dn}{dt} = \alpha \frac{16}{3} n_p^2 R^3 \dot{\gamma}.$$ (9.17)

A comparison can be made between the collision frequency or rate of orthokinetic and perikinetic flocculation, c_f and c_b, respectively:

$$\frac{c_f}{c_b} = \frac{2\alpha \eta_o R^3 \dot{\gamma}}{kT}.$$ (9.18)

If the particles are dispersed in water at a temperature of 25 °C, the ratio in equation (9.18) becomes

$$\frac{c_f}{c_b} \approx 4 \cdot 10^{17} R^3 \dot{\gamma}.$$ (9.19)

When a liquid is stirred in a beaker using a rod, the velocity gradient r shear rate is in the range 1–10 s^{-1}, with a mechanical stirrer it is about 100 s^{-1}, and at the tip of a turbine in a large reactor it can reach values as high as 1 000–10 000 s^{-1}. This means that the particle radius R must be less than 1 µm if even slow mixing can be disregarded. This shows how the effect of shear can increase the rate of aggregation.

It should be mentioned that the above analysis is for the case where there is no energy barrier, i.e. the Smoluchowski case [4]. In the presence of an energy barrier, i.e. potential limited aggregation, one must consider the contribution due to the hydrodynamic forces acting on the colliding pair [7].

9.2.2 Flocculation of sterically stabilised emulsions

For an emulsion, stabilised by adsorbed polymer, to remain deflocculated, the following criteria must be fulfilled:

(i) The droplets should be completely covered by the polymer (the amount of polymer should correspond to the plateau value). Any bare patches may cause flocculation either by van der Waals attraction (between the bare patches) or by bridging flocculation (whereby a polymer molecule will become simultaneously adsorbed on two or more drops).

(ii) The polymer should be strongly "anchored" to the droplet surfaces, to prevent any displacement during drop approach. This is particularly important for concentrated emulsions. For this purpose A–B, A–B–A block and BA$_n$ graft copolymers are the most suitable where the chain B is chosen to be highly insoluble in the medium and has a strong affinity to the surface or soluble in the oil. Examples of B groups for non-polar oils in aqueous media are polystyrene, polypropylene oxide and poly-methylmethacrylate.

(iii) The stabilising chain A should be highly soluble in the medium and strongly solvated by its molecules. Examples of A chains in aqueous media are poly(ethylelene oxide) and poly(vinyl alcohol).

(iv) The adsorbed layer thickness δ should be sufficiently large (> 5 nm) to prevent weak flocculation.

It is convenient to discuss the flocculation of sterically stabilised emulsions in three sections: (i) high molar mass terminally anchored polymer chains; (ii) low molar mass, terminally anchored chains; (iii) multi-point anchored chains, i.e. chains having a loop and train type configuration at the interface.

Flocculation of emulsions with high molar mass terminally anchored chains

In this case the mixing interaction, G_{mix}, dominates the interaction, at least at low degree of interpenetration (see Chapter 3). Here one can neglect the elastic term, G_{el}, until the surface-surface separation h is lower than δ, i.e. when the polymer layer on one interface comes in contact with the second interface itself. For high molar mass chains, the polymer adsorbed layer is sufficiently thick that one can safely neglect any contribution from the van der Waals attraction, G_A, to the total interaction. In other words, the total interaction G_T is simply given by G_{mix} that is given by

$$G_{mix} = \left(\frac{2V_2^2}{V_1} \right) v_2 \left(\frac{1}{2} - \chi \right) \left(3R + 2\delta + \frac{h}{2} \right) \tag{9.20}$$

where V_1 and V_2 are the molar volumes of solvent and polymer, respectively, v_2 is the number of chains per unit area and χ is the Flory–Huggins interaction parameter. The latter depends on the solvency of the stabilising A-chain by the medium. In a good solvent for the A chain, $\chi < 0.5$, whereas for a poor solvent for the chain $\chi > 0$. The point at which $\chi = 0.5$ is referred to as the θ-condition. The Flory–Huggins interaction parameter χ may be conveniently changed by varying the temperature, adding a non-solvent or increasing the electrolyte concentration in the external phase.

As an illustration, Fig. 9.2 shows the variation of G_{mix}, G_{el}, G_A and G_T with h as χ is increased from < 0.5 to > 0.5.

It can be seen that when $\chi > 0.5$ (i.e. the medium for the A-chains becomes worse than a θ-solvent), a significant value of G_{min} is attained, resulting in catastrophic floc-culation (sometimes referred to as incipient flocculation). With many systems good correlation between the flocculation point and the θ point is obtained. For example, the emulsion will flocculate at a temperature (referred to as the critical flocculation temperature, CFT) equal to the θ-temperature of the stabilising chain. The emulsion may flocculate at a critical volume fraction of a non-solvent (CFV), which is equal to the volume of non-solvent that brings it to a θ-solvent.

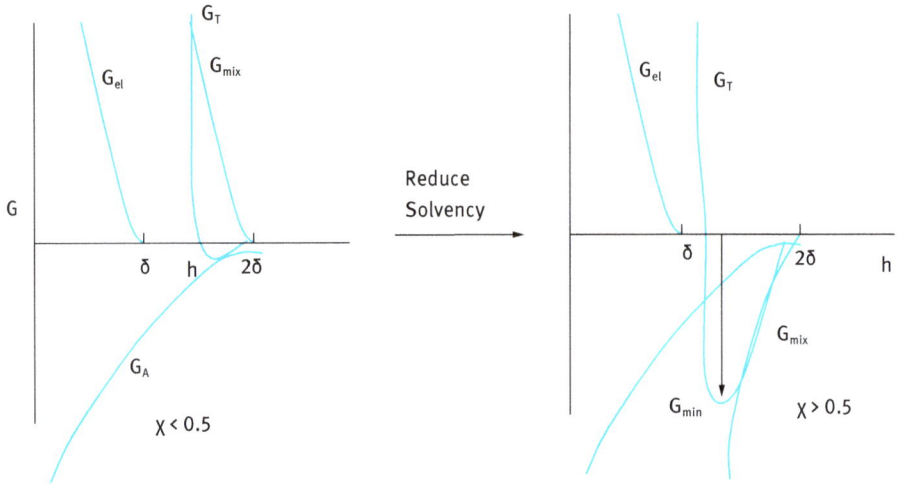

Fig. 9.2: Schematic representation of the interaction free energy separation curves for two droplets stabilised by high molar mass polymer.

It should be mentioned, however, that some emulsions flocculate on cooling and others on heating. Generally speaking (but not always), the former occurs when a non-aqueous solvent is the external phase (e.g. W/O emulsions), while the latter occurs when water is the external phase (O/W emulsions).

The correlation between the CFT and θ-temperature [1] has been demonstrated for toluene-in-water emulsions dispersed in 0.39 mol dm^{-3} MgSO$_4$ stabilised by poly-(ethylene oxide), PEO. The stability of the emulsion was assessed by monitoring the volume of cream layer. A stable emulsion shows little sign of creaming within a given period of time (2 h), whereas a flocculated emulsion shows a clear cream layer. The results are shown in Fig. 9.3.

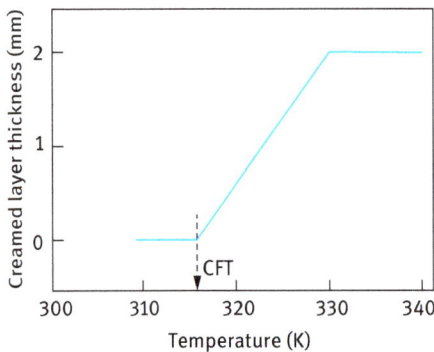

Fig. 9.3: Location of the CFT for a sterically stabilised emulsion.

Fig. 9.3 shows the clear location of the CFT, which is 318 K. The observed flocculation is reversible, i.e. the emulsion can be readily dispersed on cooling below the CFT. This suggests absence of coalescence in the cream layer.

Flocculation of emulsions with low molar mass terminally anchored chains.

In this case one cannot neglect the role played by G_{el}, since the segment concentration in the adsorbed layer is now significantly high. Since G_{el} now tends to come into play almost at the same place as G_{mix} (i.e. at $h \sim 2\delta$), the changes in G_{mix} resulting from change in solvency (i.e the χ-parameter) for the stabilising chain tend to be somewhat masked. However, changes in solvency also lead to changes in G_{el}, since a decrease in solvency leads to contraction in the thickness of the adsorbed layer (and consequently an increase in the average segment density). The result of this is that there is an increase in G_{min} with decreasing solvency, but the change is more gradual than for the high molar mass polymer, and the correlation of CFT with θ-temperature may no longer hold. A typical example of the stability-flocculation behaviour of (neutral) polystyrene latex particles with terminally attached poly(ethylene oxide) with molar mass 750 and 2000, in $0.39\,\mathrm{mol\,dm^{-3}}$ $MgSO_4$ showed that the CFT values to be significantly lower than the corresponding θ-temperature for PEO. A schematic representation of the effect of reducing solvency (e.g. by increasing temperature in the case of PEO in aqueous electrolyte solution) is shown in Fig. 9.4.

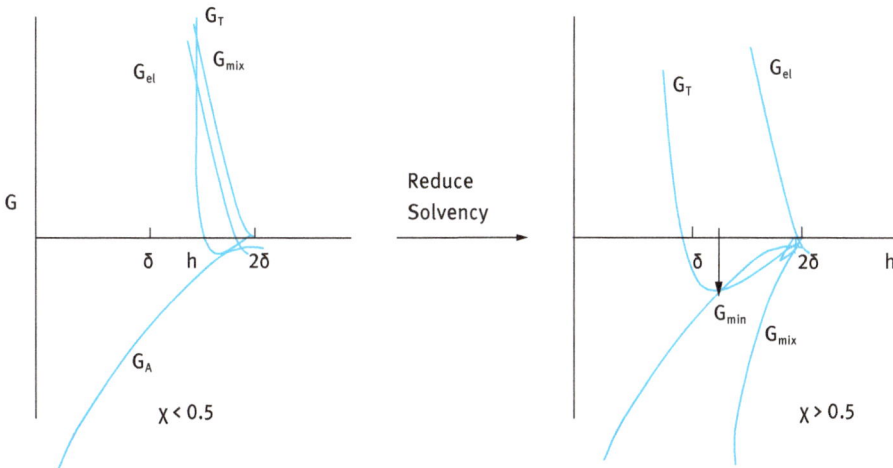

Fig. 9.4: Interaction free energy for systems stabilised by low molar mass polymer.

Flocculation of emulsions with adsorbed homopolymers/multi-point/anchored chains

Here the situation is even more complex, and it now lies somewhere between the two extremes depicted in Figures 9.3 and 9.4. The average segment density in the adsorbed layer may be quite low, and because most of the segments are likely to be in loops rather in tails, significant interpenetration without compression is unlikely. In this case G_{el} may still be important and, therefore, a close correlation of the CFT with the θ-temperature may not occur. In this case, the CFT may be much higher than the θ-temperature. A good example in this case is emulsions stabilised by partially hydrolysed poly(vinyl acetate), usually referred to as PVA [8].

9.2.3 Weak flocculation of sterically stabilised emulsions

As discussed in Chapter 3, the energy distance curve of sterically stabilised emulsions shows a shallow minimum G_{min}, at separation distances $h \sim 2\delta$, whose depth can be of the order of few kT units. The minimum depth depends on droplet radius R, Hamaker constant A and adsorbed layer thickness δ (i.e. with decrease of the molecular weight of the stabiliser). For a given R and A, G_{min} increases with decrease of δ. This is illustrated in Fig. 9.5, which shows the energy–distance curves as a function of δ/R. The smaller the value of δ/R, the larger is the value of G_{min}.

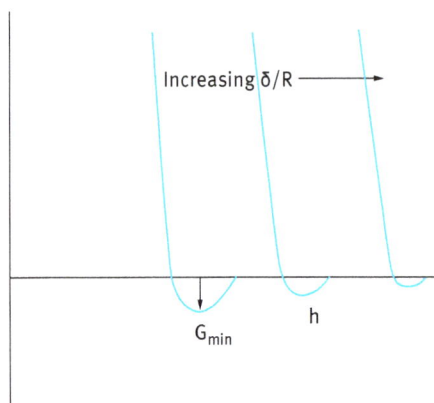

Fig. 9.5: Variation of G_{min} with δ/R.

The minimum depth required for causing weak flocculation depends on the volume fraction of the emulsion. The higher the volume fraction, the lower the minimum depth required for weak flocculation. This can be understood if one considers the free energy of flocculation consisting of two terms: an energy term determined by the depth of the minimum (G_{min}), and an entropy term determined by reduction in configurational entropy on aggregation of droplets:

$$\Delta G_{flocc} = \Delta H_{flocc} - T\Delta S_{flocc} .$$ (9.21)

With dilute emulsions, the entropy loss on flocculation is larger than with concentrated emulsions. Hence for flocculation of a dilute emulsions, a higher energy minimum is required when compared with the case with concentrated emulsions. This flocculation is weak and reversible, i.e. on shaking the container redispersion of the emulsion occurs. On standing, the dispersed droplets aggregate to form a weak "gel". This process (referred to as sol $\leftrightarrow$ gel transformation) leads to reversible time dependence of viscosity (thixotropy). On shearing the emulsion, the viscosity decreases, and when the shear is removed the viscosity is recovered.

9.2.4 Depletion flocculation

As mentioned in Chapter 8, depletion flocculation is produced by addition of "free" non-adsorbing polymer [9]. In this case, the polymer coils cannot approach the droplets to a distance Δ (determined by the radius of gyration of free polymer R_G), since the reduction of entropy on the close approach of the polymer coils is not compensated by an adsorption energy. The emulsion droplets will be surrounded by a depletion zone with thickness Δ. Above a critical volume fraction of the free polymer, φ_p^+, the polymer coils are "squeezed out" from between the droplets, and the depletion zones begin to interact. The interstices between the droplets are now free from polymer coils, and hence an osmotic pressure is exerted outside the droplet surface (the osmotic pressure outside is higher than in between the particles), resulting in weak flocculation [9]. A schematic representation of depletion flocculation is shown in Fig. 9.6

The magnitude of the depletion attraction free energy, G_{dep}, is proportional to the osmotic pressure of the polymer solution, which in turn is determined by φ_p and molecular weight M. The range of depletion attraction is proportional to the thickness of the depletion zone, Δ, which is roughly equal to the radius of gyration, R_G, of the free polymer. A simple expression for G_{dep} is [9]

$$G_{dep} = \frac{2\pi R\Delta^2}{V_1}(\mu_1 - \mu_1^0)\left(1 + \frac{2\Delta}{R}\right),$$ (9.22)

where V_1 is the molar volume of the solvent, μ_1 is the chemical potential of the solvent in the presence of free polymer with volume fraction φ_p and μ_1^0 is the chemical potential of the solvent in the absence of free polymer. $(\mu_1 - \mu_1^0)$ is proportional to the osmotic pressure of the polymer solution.

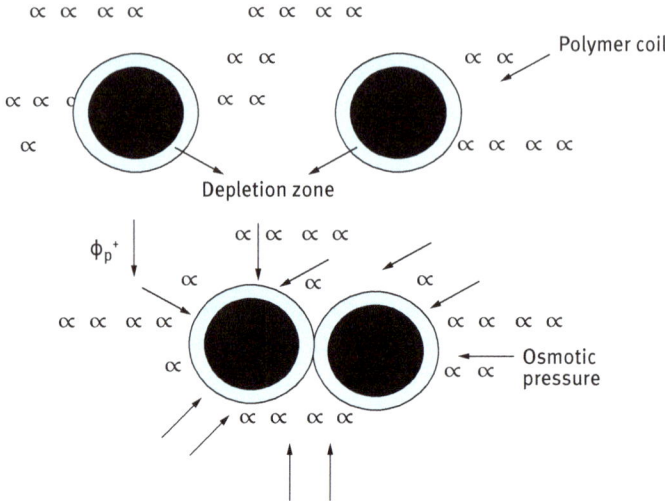

Fig. 9.6: Schematic representation of depletion flocculation.

9.2.5 Bridging flocculation by polymers and polyelectrolytes

Certain long-chain polymers may adsorb in such a way that different segments of the same polymer chain are adsorbed on different droplets, thus binding or "bridging" the droplets together, despite the electrical repulsion [10]. With polyelectrolytes of opposite charge to the droplets, another possibility exists: the droplet charge may be partly or completely neutralized by the adsorbed polyelectrolyte, thus reducing or eliminating the electrical repulsion and destabilizing the droplets.

Effective flocculants are usually linear polymers, often of high molecular weight, which may be non-ionic, anionic or cationic in character. Ionic polymers should be strictly referred to as polyelectrolytes. The most important properties are molecular weight and charge density. There are several polymeric flocculants that are based on natural products, e.g. starch and alginates, but the most commonly used flocculants are synthetic polymers and polyelectrolytes, e.g. polacrylamide and copolymers of acrylamide and a suitable cationic monomer such as dimethylaminoethyl acrylate or methacrylate. Other synthetic polymeric flocculants are poly(vinyl alcohol), poly(ethylene oxide) (non-ionic), sodium polystyrene sulphonate (anionic) and polyethyleneimine (cationic).

As mentioned above, bridging flocculation occurs because segments of a polymer chain adsorb simultaneously on different droplets, thus linking them together. Adsorption is an essential step, and this requires favourable interaction between the polymer segments and the droplets. Several types of interactions are responsible for adsorption irreversible in nature: (i) electrostatic interaction, when a polyelectrolyte adsorbs on a surface bearing oppositely charged ionic groups, e.g. adsorption of a cationic polyelectrolyte on a negative emulsion surface; (ii) hydrophobic bonding, responsible for adsorption of non-polar segments on a hydrophobic surface, e.g. partially hydrolyzed poly(vinyl acetate) (PVA) on a hydrophobic surface such as hydrocarbon oil; (iii) hydrogen bonding, as for example interaction of the amide group of polyacrylamide with hydroxyl groups on an emulsion surface; (iv) ion binding, as is the case of adsorption of anionic polyacrylamide on a negatively charged surface in the presence of Ca_2^+.

Effective bridging flocculation requires that the adsorbed polymer extends r enough from the droplet surface to attach to other droplets, and that there is sufficient free surface available for adsorption of these segments of extended chains. When excess polymer is adsorbed, the droplets can be restabilized, either because of surface saturation or by steric stabilization as discussed before. This is one explanation of the fact that an "optimum dosage" of flocculant is often found; at low concentration there is insufficient polymer to provide adequate links, and with larger amounts restabilization may occur. A schematic picture of bridging flocculation and restabilization by adsorbed polymer is given in Fig. 9.7.

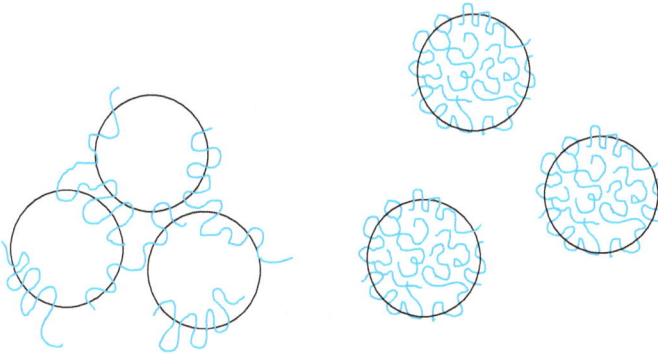

Fig. 9.7: Schematic illustration of bridging flocculation (left) and restabilization (right) by adsorbed polymer.

If the fraction of droplet surface covered by polymer is θ, then the fraction of uncovered surface is $(1 - \theta)$, and the successful bridging encounters between the droplets should be proportional to $\theta(1 - \theta)$, which has its maximum when $\theta = 0.5$. This is the well-known condition of "half-surface-coverage", which has been suggested as giving the optimum flocculation.

An important condition for bridging flocculation with charged droplets is the role of electrolyte concentration. The latter determines the extension ("thickness") of the double layer, which can reach values as high as 100 nm (in 10^{-5} mol dm^{-3} 1 : 1 electrolyte such as NaCl). For bridging flocculation to occur, the adsorbed polymer must extend far enough from the surface to a distance over which electrostatic repulsion occurs (> 100 nm in the above example). This means that at low electrolyte concentrations quite high molecular weight polymers are needed for bridging to occur. As the ionic strength is increased, the range of electrical repulsion is reduced and lower molecular weight polymers should be effective.

In many practical applications, it has been found that the most effective flocculants are polyelectrolytes with a charge opposite to that of the droplets. In aqueous media most droplets are negatively charged, and cationic polyelectrolytes such as polyethyleneimine are often necessary. With oppositely charged polyelectrolytes it is likely that adsorption occurs to give a rather flat configuration of the adsorbed chain, due to the strong electrostatic attraction between the positive ionic groups on the polymer and the negative charged sites on the droplet surface. This would probably reduce the probability of bridging contacts with other particles, especially with fairly low molecular weight polyelectrolytes with high charge density. However, the adsorption of a cationic polyelectrolyte on a negatively charged droplet will reduce the surface charge of the latter, and this charge neutralization could be an important factor in destabilizing the particles. Another mechanism for destabilization has been suggested by Gregory [10] who proposed an "electrostatic-patch" model. This applied to cases where the droplets have a fairly low density of immobile charges and the polyelectrolyte has a fairly high charge density. Under these conditions, it is not physically possible for each surface site to be neutralized by a charged segment on the polymer chain, even though the droplet may have sufficient adsorbed polyelectrolyte to achieve overall neutrality. There are then "patches" of excess positive charge, corresponding to the adsorbed polyelectrolyte chains (probably in a rather flat configuration), surrounded by areas of negative charge, representing the original particle surface. Droplets which have this "patchy" or "mosaic" type of surface charge distribution may interact in such a way that the positive and negative "patches" come into contact, giving quite strong attraction (although not as strong as in the case of bridging

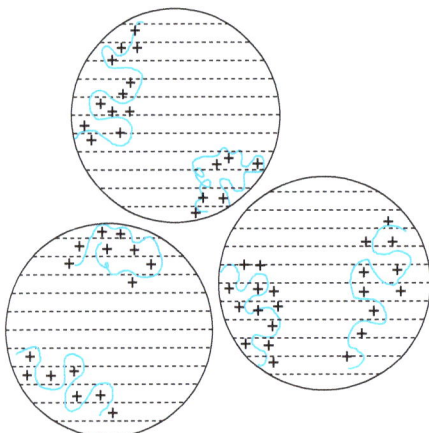

Fig. 9.8: "Electrostatic patch" model for the interaction of negatively charged droplets with adsorbed cationic polyelectrolytes.

flocculation). A schematic illustration of this type of interaction is given in Fig. 9.8. The electrostatic patch concept (which can be regarded as another form of "bridging") can explain a number of features of flocculation of negatively charged droplets with positive polyelectrolytes. These include the rather small effect of increasing the molecular weight and the effect of ionic strength on the breadth of the flocculation dosage range and the rate of flocculation at optimum dosage.

9.3 General rules for reducing (eliminating) flocculation

9.3.1 Charge stabilised emulsions, e.g. using ionic surfactants

The most important criterion is to make G_{max} as high as possible; this is achieved by three main conditions: high surface or zeta potential; low electrolyte concentration; and low valency of ions.

9.3.2 Sterically stabilised emulsions

Four main criteria are necessary: (i) Complete coverage of the droplets by the stabilising chains. (ii) Firm attachment (strong anchoring) of the chains to the droplets. This requires the chains to be insoluble in the medium and soluble in oil. However, this is incompatible with stabilisation, which requires a chain which is soluble in the medium and strongly solvated by its molecules. These conflicting requirements are solved by the use of A–B, A–B–A block or BA_n graft copolymers (B is the "anchor" chain, and A is the stabilising chain(s)). Examples for the B chains for O/W emulsions are polystyrene, polymethylmethacrylate, polypropylene oxide and alkyl polypropylene oxide. For the A chain(s), polyethylene oxide (PEO) or polyvinyl alcohol are good examples. For W/O emulsions, PEO can form the B chain, whereas the A chain(s) could be polyhydroxy stearic acid (PHS), which is strongly solvated by most oils. (iii) Thick adsorbed layers; the adsorbed layer thickness should be in the region of 5–10 nm. This means that the molecular weight of the stabilising chains could be in the region of 1000–5000. (iv) The stabilising chain should be maintained in good solvent conditions ($\chi < 0.5$) under all conditions of temperature changes on storage.

References

[1] Th. F. Tadros and B. Vincent, in: P. Becher (ed.), *Encyclopedia of Emulsion Technology*, Marcel Dekker, New York, 1983.
[2] B. V. Deryaguin and L. Landua, *Acta Physicochem. USSR* **14** (1941), 633.
[3] E. J. W. Verwey and J. T. G. Overbeek, *Theory of Stability of Lyophobic Colloids*, Elsevier, Amsterdam, 1948.

[4]   M. V. Smoluchowski, *Z. Phys. Chem.* **92** (1927), 129.

[5]   N. Fuchs, *Z. Physik* **89** (1936), 736.

[6]   H. Reerink and J. T. G. Overbeek, *Disc. faraday Soc.* **18** (1954), 74.

[7]   Tharwat Tadros, *Interfacial Phenomena, Basic Principles*, De Gruyter, Berlin/Boston, 2015.

[8]   Th. F. Tadros (ed.), *Effect of Polymers on Dispersion Stability*, Ch. 1, Academic Press, London, 1982.

[9]   A. Asakura and F. Oosawa, *J. Chem. Phys.* **22** (1954), 1235; A. Asakura and F. Oosawa, *J. Polymer Sci.* **93** (1958) , 183.

[10] J. Gregory, in: Th. F. Tadros (ed.), *Solid/Liquid Dispersions*, Academic Press, London, 1987.

10 Ostwald ripening in emulsions and its prevention

10.1 Driving force for Ostwald ripening

The driving force of Ostwald ripening is the difference in solubility between smaller and larger droplets [1]. Small droplets with radius r_1 will have higher solubility than the larger droplets with radius r_2. This can be easily recognised from the Kelvin equation [2], which relates the solubility of a particle or droplet S(r) with that of a particle or droplet with infinite radius S(∞):

$$S(r) = S(\infty) \exp \left(\frac{2\gamma V_m}{rRT} \right), \tag{10.1}$$

where γ is the solid/liquid or liquid/liquid interfacial tension, V_m is the molar volume of the disperse phase, R is the gas constant and T is the absolute temperature. The quantity $(2\gamma V_m/RT)$ has the dimension of length and is termed the characteristic length with an order of ~ 1 nm.

A schematic representation of the enhancement the solubility c(r)/c(0) with decrease of droplet size according to the Kelvin equation is shown in Fig. 10.1.

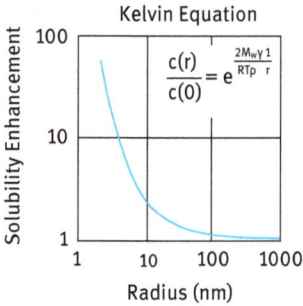

Fig. 10.1: Solubility enhancement with decrease of particle or droplet radius.

It can be seen from Fig. 10.1 that the solubility of droplets increases very rapidly with a decrease of radius, particularly when r < 100 nm. This means that a droplet with a radius of, say, 4 nm will have about 10 times the solubility enhancement compared, say, with a droplet with 10 nm radius, which has a solubility enhancement of only 2 times. Thus, with time molecular diffusion will occur between the smaller and larger droplets, with the ultimate disappearance of most of the small droplets. This results in a shift in the droplet size distribution to larger values during storage of an emulsion. This could lead to the formation of a dispersion droplet size > µm. This instability can cause severe problems, such as creaming or sedimentation, flocculation and even coalescence of the emulsion.

For two droplets with radii r_1 and r_2 ($r_1 < r_2$),

$$\frac{RT}{V_m} \ln \left[\frac{S(r_1)}{S(r_2)} \right] = 2\gamma \left[\frac{1}{r_1} - \frac{1}{r_2} \right].$$

(10.2)

Equation (10.2) is sometimes referred to as the Ostwald equation, and it shows that the larger the difference between r_1 and r_2, the higher the rate of Ostwald ripening. That is why in preparation one aims at producing a narrow size distribution.

10.2 Kinetics of Ostwald ripening

The kinetics of Ostwald ripening is described in terms of the theory developed by Lifshitz and Slesov [3] and by Wagner [4] (referred to as LSW theory). The LSW theory assumes that: (i) the mass transport is due to molecular diffusion through the continuous phase; (ii) the dispersed phase droplets are spherical and fixed in space; (iii) there is no interactions between neighbouring droplets (the droplets are separated by a distance much larger than the diameter of the droplets); (iv) the concentration of the molecularly dissolved species is constant except adjacent to the droplet boundaries.

The rate of Ostwald ripening ω is given by

$$\omega = \frac{d}{dr} \left(r_c^3 \right) = \left(\frac{8\gamma DS(\infty)V_m}{9RT} \right) f(\varphi) = \left(\frac{4DS(\infty)\alpha}{9} \right) f(\varphi),$$

(10.3)

where r_c is the radius of a particle or droplet that is neither growing nor decreasing in size, D is the diffusion coefficient of the disperse phase in the continuous phase, $f(\varphi)$ is a factor that reflects the dependence of ω on the disperse volume fraction and α is the characteristic length scale (= $2\gamma V_m/RT$).

Droplets with $r > r_c$ grow at the expense of smaller ones, while droplets with $r < r_c$ tend to disappear. The validity of the LSW theory was tested by Kabalnov et al. [5], who used 1,2 dichloroethane-in-water emulsions, whereby the droplets were fixed to the surface of a microscope slide to prevent their coalescence. The evolution of the droplet size distribution was followed as a function of time by microscopic investigations.

The LSW theory predicts that the droplet growth over time will be proportional to r_c^3. This is illustrated in Fig. 10.2 for dichloroethane-in-water emulsions.

Another consequence of the LSW theory is the prediction that the size distribution function $g(u)$ for the normalized droplet radius $u = r/r_c$ adopts a time independent form given by

$$g(u) = \frac{81 e u^2 \exp\left[1/(2u/3 - 1)\right]}{3^{21/3} (u + 3)^{7/3} (1.5 - u)^{11/3}} \quad \text{for } 0 < u \leq 1.5$$

(10.4)

and

$$g(u) = 0 \quad \text{for } u > 1.5.$$

(10.5)

A characteristic feature of the size distribution is the cut-off at $u > 1.5$.

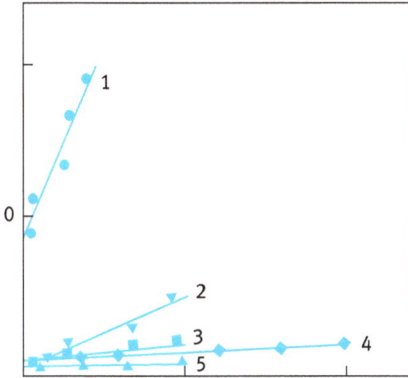

Fig. 10.2: Variation of average cube radius with time during Ostwald ripening in emulsions of (1) 1,2 dichloroethane; (2) benzene; (3) nitrobenzene; (4) toluene; (5) p-xylene.

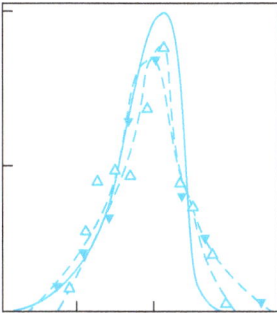

Fig. 10.3: Comparison between theoretical function g(u) (full line) and experimentally determined functions obtained for 1,2 dichloroethane droplets at time 0 (open triangles) and 300 s (inverted solid triangles).

A comparison of the experimentally determined size distribution (dichloroethane-in-water emulsions) with the theoretical calculations based on the LSW theory is shown in Fig. 10.3.

The influence of the alkyl chain length of the hydrocarbon on the Ostwald ripening rate of nanoemulsions was systematically investigated by Kabalanov et al. [6]. An increase in the alkyl chain length of the hydrocarbon used for the emulsion results in a decrease of the oil solubility. According to the LSW theory this reduction in solubility should result in a decrease of the Ostwald ripening rate. This was confirmed by the results of Kabalnov et al. [6], who showed that the Ostwald ripening rate decreases with an increase of the alkyl chain length from C_9-C_{16}. Tab. 10.1 shows the solubility of the hydrocarbon, the experimentally determined rate ω_e and the theoretical values ω_t and the ratio of ω_e/ω_t.

Although the results showed the linear dependence of the cube of the droplet radius with time in accordance with the LSW theory, the experimental rates were ~2–3 times higher than the theoretical values. The deviation between theory and experiment has been ascribed to the effect of Brownian motion [6]. The LSW theory assumes that the droplets are fixed in space, and the molecular diffusion is the only mechanism of mass transfer. For droplets undergoing Brownian motion, one must take into account the contributions of molecular and convective diffusion as predicted

Tab. 10.1: Influence of the alkyl chain length on the Ostwald ripening rate.

Hydrocarbon	$c(\infty)$ / ml ml^{-1} [a]	ω_e / cm^{-3} s^{-1}	ω_t / cm^{-3} s^{-1} [b]	$\omega_r = \omega_e/\omega_t$
C_9H_{20}	$3.1 \cdot 10^{-7}$	$6.8 \cdot 10^{-19}$	$2.9 \cdot 10^{-19}$	2.3
$C_{10}H_{22}$	$7.1 \cdot 10^{-8}$	$2.3 \cdot 10^{-19}$	$0.7 \cdot 10^{-19}$	3.3
$C_{11}H_{24}$	$2.0 \cdot 10^{-8}$	$5.6 \cdot 10^{-20}$	$2.2 \cdot 10^{-20}$	2.5
$C_{12}H_{26}$	$5.2 \cdot 10^{-9}$	$1.7 \cdot 10^{-20}$	$0.5 \cdot 10^{-20}$	3.4
$C_{13}H_{28}$	$1.4 \cdot 10^{-9}$	$4.1 \cdot 10^{-21}$	$1.6 \cdot 10^{-21}$	2.6
$C_{14}H_{30}$	$3.7 \cdot 10^{-10}$	$1.0 \cdot 10^{-21}$	$0.4 \cdot 10^{-21}$	2.5
$C_{15}H_{32}$	$9.8 \cdot 10^{-11}$	$2.3 \cdot 10^{-22}$	$1.4 \cdot 10^{-22}$	1.6
$C_{16}H_{34}$	$2.7 \cdot 10^{-11}$	$8.7 \cdot 10^{-23}$	$2.2 \cdot 10^{-23}$	4.0

a Molecular solubilities of hydrocarbons in water taken from: C. McAuliffe, *J. Phys. Chem.*, 1966, 1267.
b For theoretical calculations, the diffusion coefficients were estimated according to the Hayduk–Laudie equation (W. Hayduk and H. Laudie, *AIChE J.*, 1974, **20**, 611) and the correction coefficient $f(\varphi)$ assumed to be equal to 1.75 for $\varphi = 0.1$ (P. W. Voorhees. *J. Stat. Phys.*, 1985, **38**, 231).

by the Peclet number:

$$Pe = \frac{rv}{D},$$ (10.6)

where v is the velocity of the droplets that is approximately given by

$$v = \left(\frac{3kT}{M}\right)^{1/2},$$ (10.7)

where k is the Boltzmann constant, T is the absolute temperature and M is the mass of the droplet. For r = 100 nm, Pe = 8, indicating that the mass transfer will be accelerated with respect to that predicted by the LSW theory.

The LSW theory assumes that there are no interactions between the droplets and it is limited to low oil volume fractions. At higher volume fractions the rate of ripening depends on the interaction between the diffusion spheres of neighbouring droplets. It is expected that emulsions with higher volume fractions of oil will have a broader droplet size distribution and faster absolute growth rates than those predicted by the LSW theory. However, experimental results using high surfactant concentrations (5 %) showed the rate to be independent of the volume fraction in the range $0.01 \leq \varphi \leq 0.3$. It has been suggested that the emulsion droplets may have been screened from one another by surfactant micelles [7]. A strong dependence on volume fraction has been observed for fluorocarbon-in-water emulsions [8]. A threefold increase in ω was found when φ was increased from 0.08 to 0.52.

It has been suggested that micelles play a role in facilitating the mass transfer between emulsion droplets by acting as carriers of oil molecules [9]. Three mechanisms were suggested: (i) oil molecules are transferred via direct droplet/micelle collisions; (ii) oil molecules exit the oil droplet and are trapped by micelles in the immediate vicinity of the droplet; (iii) oil molecules exit the oil droplet collectively with a large number of surfactant molecules to form a micelle.

In mechanism (i) the micellar contribution to the rate of mass transfer is directly proportional to the number of droplet/micelle collision, i.e. to the volume fraction of micelles in solution. In this case the molecular solubility of the oil in the LSW equation is replaced by the micellar solubility, which is much higher. Large increases in the rate of mass transfer would be expected with an increase in micelle concentration. Numerous studies indicate, however, that the presence of micelles affects the mass transfer to only a small extent [10]. Results were obtained for decane-in-water emulsions using sodium dodecyl sulphate (SDS) as emulsifier at concentrations above the critical micelle concentration (CMC). This is illustrated in Fig. 10.4 which shows plots of $(d_{inst}/d_{inst}^o)^3$ where d_{inst} is the diameter at time t and d_{inst}^o the diameter at time 0) as a function of time. The results showed only a two-fold increase in ω above the CMC. This result is consistent with many other studies which showed an increase in the mass transfer of only 2–5 times with increasing micelle concentration. The lack of a strong dependence of mass transfer on micelle concentration for ionic surfactants may result from electrostatic repulsion between the emulsion droplets and micelles, which provide a high energy barrier preventing droplet/micelle collision.

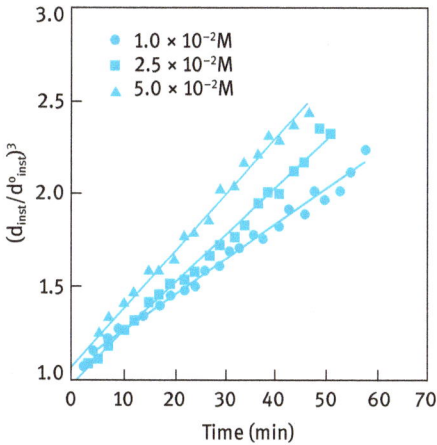

Fig. 10.4: Variation of $(d_{inst}/d_{inst}^o)^3$ with time for decane-in-water emulsions for different SDS concentrations above the CMC.

In mechanism (ii), a micelle in the vicinity of an emulsion droplet rapidly takes up dissolved oil from the continuous phase. This "swollen" micelle diffuses to another droplet, where the oil is redeposited. Such a mechanism would be expected to result in an increase in the mass transfer over and above that expected from the LSW theory by a factor φ given by the following equation:

$$\varphi = 1 + \frac{\varphi_s \Gamma D_m}{D} = 1 + \frac{\chi^{eq} D_m}{c^{eq} D}, \tag{10.8}$$

where φ_s is the volume fraction of micelles in solution, $\chi^{eq} = \varphi_s c_m^{eq}$ is the net oil solubility in the micelle per unit volume of micellar solution reduced by the density of the solute, $\Gamma = c_m^{eq}/c^{eq} \sim 10^6–10^{11}$ is the partition coefficient for the oil between the micelle and bulk aqueous phase at the saturation point, D_m is the micellar diffusivity ($\sim 10^{-6}–10^{-7}$ cm^2 s^{-1}). For a decane-water nanoemulsion in the presence of 0.1 mol dm^{-3} SDS, equation (10.8) predicts an increase in the rare of ripening by three orders of magnitude, in sharp contrast to the experimental results.

To account for the discrepancy between the theory and experiments in the presence of surfactant micelles, Kabalanov [11] considered the kinetics of micellar solubilisation, and he proposed that the rate of oil monomer exchange between the oil droplets and the micelles is slow and rate-determining. Thus, at low micellar concentration, only a small proportion of the micelles are able to rapidly solublise the oil. This leads to a small but measurable increase in the Ostwald ripening rate with micellar concentration. Taylor and Ottewill [12] proposed that micellar dynamics may also be important. According to Aniansson et al. [13], micellar growth occurs in a step-wise fashion and is characterised by two relaxation times, τ_1 and τ_2. The short relaxation time τ_1 is related to the transfer of monomers in and out of the micelles, while the long relaxation time τ_2 is the time required for break-up and reformation of the micelle. At low SDS (0.05 mol dm^{-3}) concentration $\tau_2 = \sim 0.01$ s, whereas at higher SDS concentration (0.2 mol dm^{-3}) $\tau_2 = \sim 6$ s. Taylor and Ottewill [12] suggested that at low SDS concentration τ_2 may be fast enough to have an effect on the Ostwald ripening rate, but at 5 % SDS τ_2 may be as long as 1000 s (taking into account the effect of solubilisation on τ_2), which is too long to have a significant effect on the Ostwald ripening rate.

When using non-ionic surfactant micelles, larger increases in the Ostwald ripening rate might be expected, due to the larger solubilisation capacities of the non-ionic surfactant micelles and absence of electrostatic repulsion between the nanoemulsion droplets and the uncharged micelles. This was confirmed by Weiss et al. [14] who found a large increase of the Ostwald rpinening rate in tetradecane-in-water emulsions in the presence of Tween 20 micelles.

10.3 Reduction of Ostwald ripening

10.3.1 Addition of a small proportion of highly insoluble oil

Huguchi and Misra [15] suggested that by addition of a second disperse phase which is virtually insoluble in the continuous phase, such as squalane, can significantly reduce the Ostwald ripening rate. In this case, significant partitioning between different droplets is predicted, with the component having the low solubility in the continuous phase (e.g. squalane) being expected to be concentrated in the smaller droplets. During Ostwald ripening in a two-component disperse system, equilibrium is established when the difference in chemical potential between different sized droplets, which results from curvature effects, is balanced by the difference in chemical potential resulting from partitioning of the two components. Huguchi and Misra [15] derived the following expression for the equilibrium condition, wherein the excess chemical potential of the medium soluble component, $\Delta\mu_1$, is equal for all of the droplets in a polydisperse medium:

$$\frac{\Delta\mu_i}{RT} = \left(\frac{a_1}{r_{eq}}\right) + \ln\left(1 - X_{eq2}\right) = \left(\frac{a_1}{r_{eq}}\right) - X_{02}\left(\frac{r_o}{r_{eq}}\right)^3 = \text{const.}, \qquad (10.9)$$

where $\Delta\mu_1 = \mu_1 - \mu_1^*$ is the excess chemical potential of the first component with respect to the state μ_1^* when the radius $r = \infty$ and $X_{02} = 0$, r_o and r_{eq} are the radii of an arbitrary drop under initial and equilibrium conditions respectively, X_{02} and X_{eq2} are the initial and equilibrium mole fractions of the medium insoluble component 2, a_1 is the characteristic length scale of the medium soluble component 1.

The equilibrium determined by equation (10.9) is stable if the derivative $\partial\Delta\mu_1/\partial r_{eq}$ is greater than zero for all the droplets in a polydisperse system. Based on this analysis, Kabalanov et al. [16] derived the following criterion:

$$X_{02} > \frac{2a_1}{3d_o}, \qquad (10.10)$$

where d_o is the initial droplet diameter. If the stability criterion is met for all droplets, two patterns of growth will result, depending on the solubility characteristic of the secondary component. If the secondary component has zero solubility in the continuous phase, then the size distribution will not deviate significantly from the initial one, and the growth rate will be equal to zero. In the case of limited solubility of the secondary component, the distribution is governed by rules similar to the LSW theory, i.e. the distribution function is time-variant. In this case, the Ostwald ripening rate ω_{mix} will be a mixture growth rate that is approximately given by the following equation [16]:

$$\omega_{mix} = \left(\frac{\varphi_1}{\omega_1} + \frac{\varphi_2}{\omega_2}\right)^{-1}, \qquad (10.11)$$

where φ_1 is the volume fraction of the medium soluble component and φ_2 is the volume fraction of the medium insoluble component, respectively.

If the stability criterion is not met, a bimodal size distribution is predicted to emerge from the initially monomodal one. Since the chemical potential of the soluble component is predicted to be constant for all the droplets, it is also possible to derive the following equation for the quasi-equilibrium component 1:

$$X_{02} + \frac{2a_1}{d} = \text{const.}, \qquad (10.12)$$

where d is the diameter at time t.

Kabalanov et al. [17] studied the effect of addition of hexadecane to a hexane-in-water nanoemulsion. Hexadecane, which is less soluble than hexane, was studied at three levels $X_{02} = 0.001, 0.01$ and 0.1. For the higher mole fraction of hexade-cane, namely 0.01 and 0.1, the emulsion had the a physical appearance to that of an emulsion containing only hexadecane, and the Ostwald ripening rate was reliably pre-dicted by equation (10.11). However, the emulsion with $X_{02} = 0.001$ quickly separated into two layers: a sedimented layer with a droplet size of ca. $5\,\mu m$ and a dispersed pop-ulation of submicron droplets (i.e. a bimodal distribution). Since the stability criterion was not met for this low volume fraction of hexadecane, the observed bimodal distri-bution of droplets is predictable.

10.3.2 Modification of the interfacial layer for reduction of Ostwald ripening

According to LSW theory, the Ostwald ripening rate ω is directly proportional to the interfacial tension γ. Thus by reducing γ, ω is reduced. This could be confirmed by measuring ω as a function of SDS concentration for decane-in-water emulsion [10] be-low the critical micelle concentration (cmc). Below the cmc, γ shows a linear decrease with increase in log [SDS] concentration. The results are summarised in Tab. 10.2.

Tab. 10.2: Variation of Ostwald ripening rate with SDS concentration for decane-in-water emulsions.

[SDS] Concentration / mol dm^{-3}	ω / cm^3 s^{-1}
0.0	$2.50 \cdot 10^{-18}$
$1.0 \cdot 10^{-4}$	$4.62 \cdot 10^{-19}$
$5.0 \cdot 10^{-4}$	$4.17 \cdot 10^{-19}$
$1.0 \cdot 10^{-3}$	$3.68 \cdot 10^{-19}$
$5.0 \cdot 10^{-3}$	$2.13 \cdot 10^{-19}$

cmc of SDS $= 8.0 \cdot 10^{-3}$

Several other mechanisms have been suggested to account for the reduction of Ost-wald ripening rate by modification of the interfacial layer. For example, Walstra [18] suggested that emulsions could be effectively stabilised against Ostwald ripening by the use of surfactants which are strongly adsorbed at the interface and which do not desorb during the Ostwald ripening process. In this case the increase in interfacial dilational modulus ε and decreases in interfacial tension γ would be observed for the shrinking droplets. Eventually the difference in ε and γ between droplets would balance the difference in capillary pressure (i.e. curvature effects), leading to a quasi-equilibrium state. In this case, emulsifiers with low solubilities in the continuous phase such as proteins would be preferred. Long-chain phospholipids with a very low solubility (cmc $\sim 10^{-10}$ mol dm^{-3}) are also effective in reducing Ostwald ripening of some emulsions. The phospholipid would have to have a solubility in water about three orders of magnitude lower than the oil [19].

10.4 Influence of initial droplet size of emulsions on the Ostwald ripening rate

The influence of initial droplet size on Ostwald ripening can be realised by considering the droplet size dependence of the characteristic time, τ_{OR}:

$$\tau_{OR} \approx \frac{r^3}{\alpha S(\infty)D} \approx \frac{r^3}{\omega}. \tag{10.13}$$

The values of τ_{OR} when $r = 100$ nm are given in Tab. 10.3 for a series of hydrocarbons with increasing chain lengths, which clearly shows the reduction in Ostwald ripening rate with the increase in chain length (due to reduction in solubility $S(\infty)$). The dramatic dependence of τ_{OR} on the increase in the chain length is apparent. The characteristic time shows a large dependence on the initial average radius, as illustrated in Fig. 10.5 for a series of emulsions. It can be seen that the Ostwald ripening rate can be extremely rapid for small droplet sizes, thereby providing a key component in determining initial droplet size. For example, it is not likely that droplets with radii less than 100 nm will be observed for decane-in-water nanoemulsions since the droplets will ripen to this size on the time scale of few minutes. This was confirmed by Kabalanov et al. [5], who noted large differences in initial droplet size for hydrocarbon-in-water emulsions as the chain length of the hydrocarbon was decreased. For example, nonane-in-water nano-emulsions had an initial droplet size of 178 nm, decane-in-water nano-emulsions had a size of 124 nm and undecane-in-water nano-emulsions had a size of 88 nm. It is clear from Fig. 10.5 that the driving force for Ostwald ripening decreases dramatically with increasing droplet size.

Tab. 10.3: Characteristic time for Ostwald ripening in hydrocarbon-in-water nanoemulsions stabilized by 0.1 mol dm^{-3} SDS.

Hydrocarbon	ω_e / cm^3 s^{-1}	$\tau_{OR} \sim (r^3/\omega_e)$
C_9H_{20}	$6.8 \cdot 10^{-19}$	25 min
$C_{10}H_{22}$	$2.3 \cdot 10^{-19}$	73 min
$C_{11}H_{24}$	$5.6 \cdot 10^{-20}$	5 h
$C_{12}H_{26}$	$1.7 \cdot 10^{-20}$	16 h
$C_{13}H_{28}$	$4.1 \cdot 10^{-21}$	3 d
$C_{14}H_{30}$	$1.0 \cdot 10^{-21}$	12 d
$C_{15}H_{32}$	$2.3 \cdot 10^{-22}$	50 d
$C_{16}H_{34}$	$8.7 \cdot 10^{-23}$	133 d

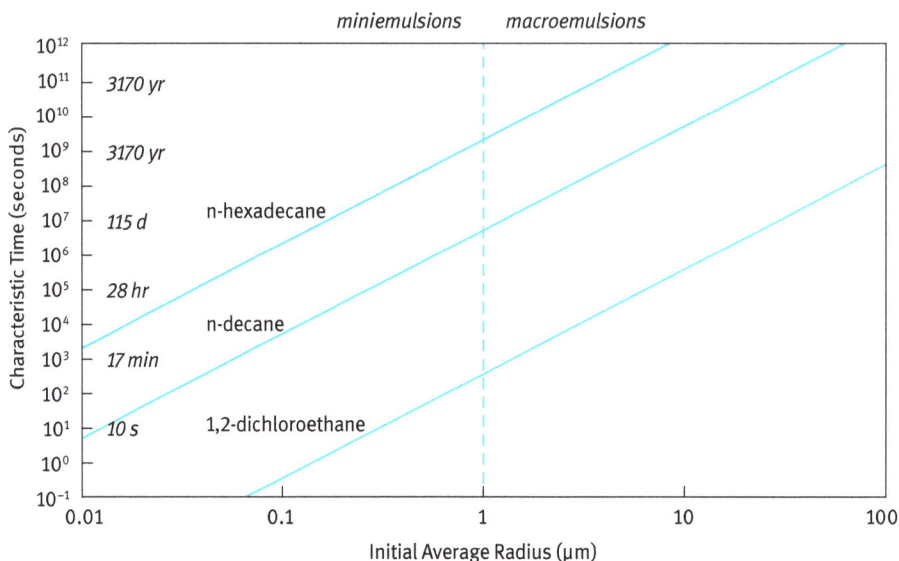

Fig. 10.5: Characteristic time for Ostwald ripening vs droplet size.

References

[1]   J. G. Weers, Molecular Diffusion in Emulsions and Emulsion Mixtures, in: B. P. Banks (ed.), *Modern Aspects of Emulsion Science*, Royal Society of Chemistry Publication, Cambridge, UK, 1998.

[2]   W. Thompson (Lord Kelvin), Phil. Mag., **42**, 448 (1871).

[3]   I. M. Lifshitz and V. V. Slesov, *Sov. Phys. JETP* **35**, 331 (1959).

[4]   C. Wagner, *Z. Electrochem.* **35** (1961) , 581.

[5]   A. S. Kabalanov and E. D. Shchukin, *Adv. Colloid Interface Sci.* **38** (1992), 69.

[6]   A. S. Kabalanov, K. N. Makarov, A. V. Pertsov and E. D. Shchukin, *J. Colloid Interface Sci.* **138** (1990), 98.

[7]   P. Taylor, *Colloids and Surfaces A* **99** (1995), 175.

[8]   Y. Ni, T. J. Pelura, T. A. Sklenar, R. A. Kinner and D. Song, *Art. Cells Blood Subs. Immob. Biorech.* **22** (1994), 1307.

[9]   S. Karaboni, N. M. van Os, K. Esselink and P. A. J. Hilbers, *Langmuir* **9** (1993), 1175.

[10] J. Soma and K. D. Papadadopoulos, *J. Colloid Interface Sci.* **181** (1996), 225.

[11] A. S. Kabalanov, *Langmuir* **10** (1994), 680.

[12] P. Taylor and R. H. Ottewill, *Colloids and Surfaces A*, **88** (1994), 303.

[13] E. A. G. Aniansson, S. N. Wall, M. Almegren, H. Hoffmann, I. Kielmann, W. Ulbricht, R. Zana, J. Lang and C. Tondre, *J. Phys. Chem.* **80** (1976), 905.

[14] J. Weiss, J. N. Coupland, D. Brathwaite and D. J. McClemments, *Colloids and Surfaces A* **121** (1997), 53.

[15] W. I. Higuchi and J. Misra, *J. Pharm. Sci.* **51** (1962) , 459.

[16] A. S. Kabalanov, A. V. Pertsov and E. D. Shchukin, *Colloids and Surfaces* **24** (1987), 19.

[17] A. S. Kabalanov, A. V. Pertsov, Y. D. Aprosin and E. D. Shchukin, *Kolloid Zh.* **47** (1095), 1048.

[18] P. Walstra, in: P. Becher (ed.), *Encyclopedia of Emulsion Technology*, vol. 4, Marcel Dekker, New York, 1996.

[19] A. Kabalanov, J. Weers, P. Arlauskas and T. Tarara, *Langnuir* **11** (1995), 2966.

11 Emulsion coalescence and its prevention

11.1 Introduction

When two emulsion droplets come in close contact in a floc or creamed layer or dur-ing Brownian diffusion, a thin liquid film or lamella forms between them [1]. This is illustrated in Fig. 11.1.

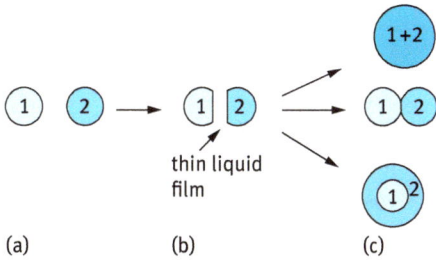

Fig. 11.1: Droplet coalescence, adhesion and engulfment.

Coalescence results from the rupture of this film, as illustrated in Figure 1c at the top. If the film cannot be ruptured, adhesion (Fig. 11.1 (c) middle) or engulfment (Fig. 11.1 (c) bottom) may occur. Film rupture usually commences at a specified "spot" in the lamella, arising from thinning in that region. This is illustrated in Fig. 11.2, where the liquid surfaces undergo some fluctuations forming surface waves. The surface waves may grow in amplitude, and the apices may join as a result of the strong van der Waals attraction (at the apex, the film thickness is the smallest). The same applies if the film thins to a small value (critical thickness for coalescence). In order to under-stand the behaviour of these films, one has to consider two aspects of their physics: (i) the nature of the forces acting across the film; these determine whether the film is thermodynamically stable, metastable or unstable; (ii) the kinetic aspects associated with local (thermal or mechanical) fluctuations in film thickness.

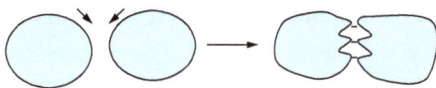

Fig. 11.2: Schematic representation of surface fluctuations.

11.2 Forces across liquid films

Fig. 11.3 shows the general features of the lamella between two droplets of phase α in a continuous phase β. The film consists of two flat parallel interfaces separated by a distance b. At the end of the film there is a border or transition region, where the inter-faces have a high curvature, i.e. compared to the curvature of the droplets themselves.

Eventually at larger values of b (effectively beyond the range of forces operating across the film) the curvature decreases to that of the droplets themselves, i.e. becomes effectively flat, on the scale considered here, for droplets in the 1 μm region. One may define a macroscopic contact angle θ, as shown in Fig. 11.3.

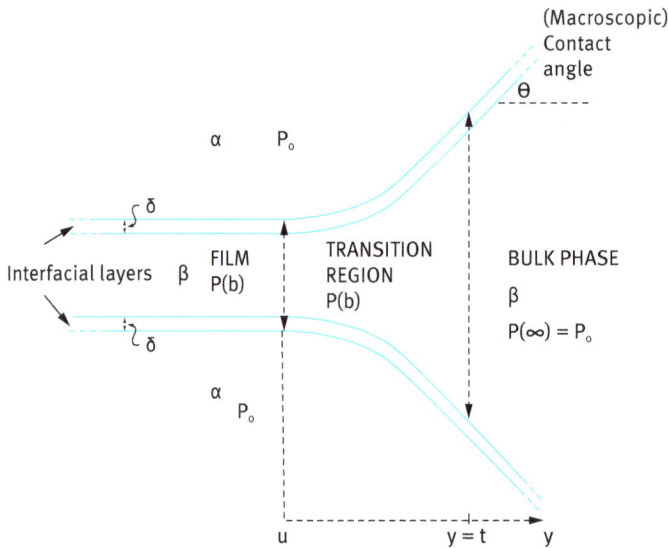

Fig. 11.3: Schematic representation of the thin film and border regions between two liquid droplets (α) in a continuous phase (β).

In considering the forces acting across the film, two regions of separation are of interest: (i) b > 2δ, where δ is the film thickness, where the forces acting are long-range van der Waals forces and electrical double layer interactions as described by the DLVO (Deyaguin–Landau–Verwey–Overbeek) theory [2, 3], schematically shown in Fig. 11.4. This shows a secondary minimum, an energy maximum and a primary minimum. When the film is sitting in either the primary or secondary minimum, the net force on the film is zero (i.e. $dG/dh = 0$). These two metastable states correspond with the so-call Newton black films, respectively.

(ii) When the film is in the primary minimum and b < 2δ, steric interactions come into play, as discussed in Chapter 3. In this case G increases very sharply with decrease of b, and for film rupture to occur these steric interactions must breakdown. Two main approaches were considered to analyse the stability of thin films in terms of the relevant interactions. The first approach was considered by Deryaguin [4], who introduced the concept of disjoining pressure. The second approach considered the interfacial tension of the film that could be related to the tangential pressure across the interface [5]. Both approaches are described below.

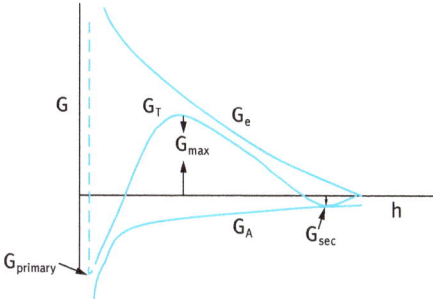

Fig. 11.4: Energy–distance curves according to the DLVO theory [2, 3].

11.2.1 Disjoining pressure approach

Deryaguin [4] suggested that a "Disjoining Pressure" $\pi(h)$ is produced in the film which balances the excess normal pressure:

$$\pi(b) = P(b) - P_o , \tag{11.1}$$

where $P(b)$ is the pressure of a film with thickness b and P_o is the pressure of a sufficiently thick film such that the net interaction free energy is zero.

$\pi(b)$ may be equated to the net force (or energy) per unit area acting across the film:

$$\pi(b) = -\frac{dG_T}{db} , \tag{11.2}$$

where G_T is the total interaction energy in the film.

$\pi(b)$ is made of three contributions due to electrostatic repulsion (π_E), Steric repulsion (π_s) and van der Waals attraction (π_A):

$$\pi(b) = \pi_E + \pi_s + \pi_A . \tag{11.3}$$

To produce a stable film $\pi_E + \pi_s > \pi_A$, and this is the driving force for prevention of coalescence which can be achieved by two mechanisms and their combination: (i) increased repulsion both electrostatic and steric; (ii) dampening of the fluctuation by enhancing the Gibbs elasticity. In general smaller droplets are less susceptible to surface fluctuations and hence coalescence is reduced.

11.2.2 Interfacial tension of liquid films

The interfacial tension $\gamma(b)$ can be related to the variation in the tangential pressure tensor p_t across an interface. There will be a similar variation in P_t across the liquid film at some thickness b, as shown schematically in Fig. 11.5.

One can define an interfacial tension for the whole film:

$$\gamma(b) = \int_0^b (P_t - P_0) \, dx . \tag{11.4}$$

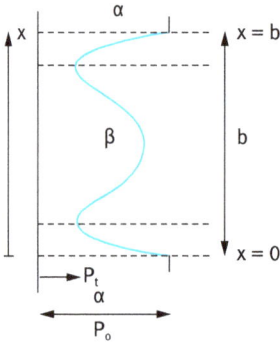

Fig. 11.5: The variation in tangential pressure P_t across a thin film.

By choosing some dividing plane in the middle of the film (conveniently at $x = b/2$, for a symmetrical film), one can divide $\gamma(b)$ into two contributions, one from the upper interface and one from the lower interface:

$$\gamma(b) = \gamma^{\alpha\beta}(b) + \gamma^{\beta\alpha}(b) = 2\gamma^{\alpha\beta}(b) .\qquad(11.5)$$

Note that $\gamma^{\alpha\beta}(b) \neq \gamma^{\alpha\beta}(\infty)$, the interfacial tension of an isolated $\alpha\beta$ interface, i.e. between the bulk liquids α or β, for an emulsion, in the region of the interface of the droplet far away from the contact zone. $\gamma(b)$ is related to $G_i(b)$ and $\pi(b)$ through the relations:

$$\gamma(b) = \gamma(\infty) - G_i(b) ,\qquad(11.6)$$

$$\gamma(b) = \gamma(\infty) + \int_{\infty}^{b} \pi(b)\, db .\qquad(11.7)$$

Equation (11.7) is obtained by combining equations (11.6) and (11.2).

11.3 Film rupture

Film rupture is a nonequilibrium effect and is associated with local thermal or mechanical fluctuations in the film thickness b. A necessary condition for rupture to occur, i.e. for a spontaneous fluctuation to occur, is that

$$\frac{d\pi_A}{db} > \frac{d\pi_E}{db} .\qquad(11.8)$$

However, this would assume that at a given value of b there are no changes in $\gamma(b)$ fluctuations. This is not so, since a local fluctuation is necessarily accompanied by a local increase in the interfacial area, resulting in a decrease in surfactant or polymer adsorption in that region, and, therefore, a local rise in the interfacial tension. This effect (referred to as the Gibbs–Marangoni effect; see Chapter 5) opposes the fluctuation. Equation (11.8) has to be modified by including a term at the right hand side to take into account the fluctuation effect in the local interfacial tension:

$$\frac{d\pi_A}{db} > \frac{d\pi_E}{db} + \frac{d\pi_y}{db}. \tag{11.9}$$

As a film thins locally due to fluctuations, if the conditions met by equation (11.9) is met at a critical thickness, b_{cr}, then the film becomes unstable, the fluctuation "growth" leading to rupture. Scheludko [5] introduce the concept of a critical thickness, and he derived the following equation for the critical thickness:

$$b_{cr} = \left(\frac{A\pi}{32K^2\gamma_o}\right)^{1/4}, \tag{11.10}$$

where A is the net Hamaker constant of the film, K is the wave number of the fluctuation and γ_o [= $\gamma(\infty)$] the interfacial tension of the isolated liquid/liquid interface. K depends on the radius of the (assumed) circular film zone.

Vrij [6] derived alternative expressions for b_{cr}. For large thicknesses where $\pi_A \ll \pi_y$,

$$b_{cr} = 0.268\left(\frac{A^2R^2}{\gamma_o\pi_yf}\right). \tag{11.11}$$

For small thicknesses, where $\pi_y \ll \pi_{yA}$,

$$b_{cr} = 0.22\left(\frac{AR^2}{\gamma f}\right)^{1/4}, \tag{11.12}$$

where f is a factor that depends on b.

The Scheludko and Vrij equations (11.10) and (11.12) have the same form at small thicknesses. Equation (11.12) predicts that when $\gamma \to 0$, the film should spontaneously rupture at large b values. This is certainly not observed, since when $\gamma \to 0$ the emulsion becomes highly stable. Also equation (11.12) predicts that as $R \to 0$, $b_{cr} \to 0$, i.e, very small droplets should never rupture. Experiments on aqueous foam films suggest that one observes finite values for b_{cr} as $R \to 0$. Experiments on emulsion droplets showed no changes in b_{cr} with a change in the size of the contact area. This is because the lamella formed between two oil droplets at non-equilibrium separations does not have the idealised, planer interface depicted in Fig. 11.3. Rather they have a "dimple" structure, as illustrated in Fig. 11.6.

This "dimple" structure does not arise from fluctuations, but rather from an effect produced by the draining of the solution from the film region and associated with

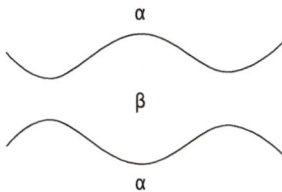

Fig. 11.6: Schematic representation of the "dimple" between two emulsion droplets.

hydrodynamic effects. The thinnest region of the film occurs at the periphery of the contact zone, and rupture tends to occur here. With polymer-stabilised films, dimpling is far less marked, due to the increased rigidity of the interface. The dimpling effect also accounts for the fact that the interfacial tension γ_o seems to have little effect on film rupture in emulsion systems.

11.4 Rate of coalescence between droplets

Van den Tempel [7] derived an expression for the coalescence rate of emulsion droplets by assuming the rate to be proportional to the number of contact points between the droplets in an aggregate. Both flocculation and coalescence are taken into account simultaneously. The average number of primary droplets n_a in an aggregate at time t is given by Smoluchowski theory (see Chapter 9). The number of droplets n which have not yet combined into aggregates at time t is given by

$$n = \frac{n_o}{(1 + kn_ot)^2} , \tag{11.13}$$

where n_o is the initial number of droplets.

The number of aggregates n_v is given by

$$n_v = \frac{kn_o^2 t}{(1 + kn_ot)^2} . \tag{11.14}$$

The total number of primary droplets in all aggregates is given by

$$n_o - n_t = n_o \left[1 - \frac{1}{(1 + kn_ot)^2} \right] . \tag{11.15}$$

Hence,

$$n_a = \frac{(n_o - n_t)}{n_o} = 2 + an_ot , \tag{11.16}$$

where a now denotes the rate of flocculation.

If m is the number of separate droplets existing in an aggregate, then m < n_a, as some coalescence will have occurred; m will be only slightly lower than n_a if coalescence is slow, whereas m → 1 if coalescence is very rapid. The rate of coalescence is then proportional to m − 1, i.e. the number of contacts between droplets in an aggregate. In a small aggregate, van den Tempel [6] observed that in sufficiently dilute emulsions small aggregates generally contain one large droplet together with one or two small ones and are built up linearly. Thus n_v decreases in direct proportion to m − 1, whereas m increases at the same time by adhesion to other droplets. The rate of increase caused by flocculation is given (following equation (11.16)) by

$$\frac{dm}{dt} = an_o - K(m - 1) , \tag{11.17}$$

where K is the rate of coalescence.

Integrating equation (11.17), for the boundary conditions m = 2 for t = 0,

$$m - 1 = \frac{an_o}{K} + \left(1 - \frac{an_o}{K}\right) \exp\left(-Kt\right). \tag{11.18}$$

The total number of droplets, whether flocculated or not, in a coagulating emulsion at time t is obtained by adding the number of unreacted primary droplets to the number of droplets in aggregates:

$$n = n_t + n_v m = \frac{n_v}{1 + kn_o t} + \frac{kn_o^2 t}{(1 + kn_o t)^2}\left[\frac{kn_o}{K} + \left(1 - \frac{kn_o}{K}\right)\exp\left(-Kt\right)\right]. \tag{11.19}$$

The first term on the right-hand side of equation (11.19) represents the number of droplets which would have been present if each droplet has been counted as a single droplet. The second term gives the number of droplets which arise when the composition of the aggregates is taken into account. In the limiting case $K \to \infty$, the second term on the right hand side of equation (11.19) is equal to zero, and the equation reduces to the Smoluchowski equation (see Chapter 9). On the other hand, if K = 0, i.e. no coalescence occurs, $n = n_o$ for all values of t. For the case $0 < K < \infty$, the effect of a change in the droplet number concentration on the rate of flocculation is given by equation (11.19). This clearly shows that the change in droplet number concentration with time depends on the initial droplet number concentration n_o. This illustrates the difference between emulsions and suspensions. In the latter case, the rate of increase of $1/n$ with time is independent of the particle number concentration. Some calculations, using reasonable values for the rate of flocculation (denoted by a equivalent to k in equation (11.19)) and rate of coalescence K are shown in Fig. 11.7 for various values of n_o. It is clear that the rate of increase of $1/n$ (or decrease in the droplet number concentration) with t rises more rapidly as n_o decreases. Van den Tempel plotted $\Delta(1/n)$, i.e. the decrease of droplet number concentration after 5 min, vs the initial droplet number concentration n_o for two values of K and k (or a). The results are shown in Fig. 11.8, which clearly shows that the rate of flocculation, as measured by the value of $1/n$, does not change significantly with n_o, neither for dilute nor for concentrated emulsions. However in the region where $kn_o K$ is of the order of unity, the rate of flocculation decreases sharply with increase of n_o.

To simplify equation (11.19), van den Tempel made three approximations:

(i) In a flocculating concentrated emulsion $kn_o \gg K$. In most real systems K is generally much smaller than unity, and $kn_o \geq 1$ is sufficient to satisfy this condition. Thus, kn_o rapidly becomes larger than unity, and the contribution from unreacted primary droplets may be neglected. In this case equation (11.19) reduces to

$$n = \frac{kn_o^2 t}{(1 + kn_o t)^2} \frac{kn_o}{K}\left[1 - \exp\left(-kt\right)\right]. \tag{11.20}$$

Since $kn_o t \gg 1$, then $1 + kn_o t \sim kn_o t$, so that

$$n = \frac{n_o}{Kt}\left[1 - \exp\left(-kt\right)\right]. \tag{11.21}$$

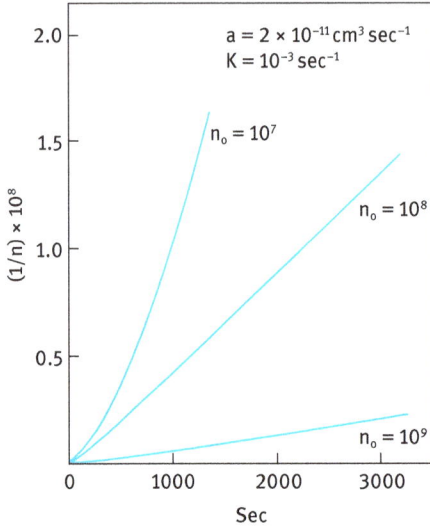

Fig. 11.7: Variation of 1/n, with t at various n_o values.

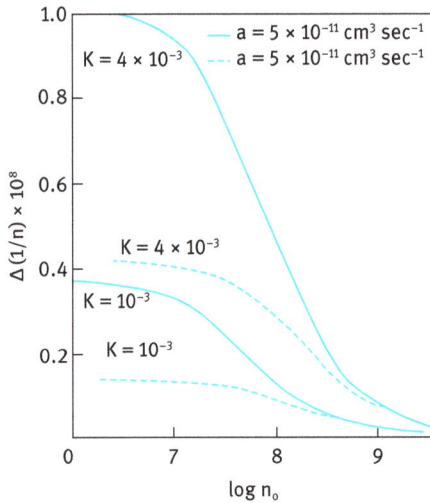

Fig. 11.8: Plot of $\Delta(1/n)$ vs log n_o for two values of K and a (or k).

This means that the rate of coalescence no longer depends on the rate of flocculation for concentrated emulsions. Van den Tempel calculated the change in droplet number concentration with time for concentrated ($n_o > 10^{10}$ cm^{-3}) and dilute emulsion ($n_o = 10^9$ cm^{-3}) and for values of k $= 5 \cdot 10^{-11}$ cm^3 s^{-1} and K $= 10^3$ s^{-1}; the results are shown in Fig. 11.9.

For concentrated emulsions, equations (11.20)–(11.22) yield similar results, whereas for dilute emulsions, equation (11.22) gives rise to a serious deviation at values of t less than 1000 s. Moreover, the droplet number concentration is found to decrease approximately exponentially with time, until Kt becomes large compared to unity.

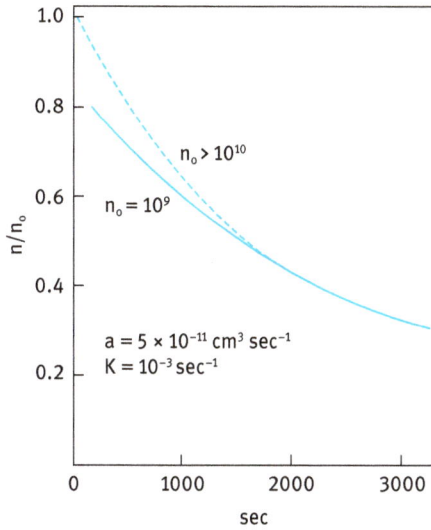

Fig. 11.9: Change in droplet number concentration with time for dilute and concentrated emulsions.

Another limitation for the application of equations (11.20)–(11.22) is the assumption made in their derivation, where the number of contact points between m droplets was taken to be m − 1. This is certainly not the case for concentrated emulsions, where the aggregates contain a large number of contacts. In a closely packed aggregate of spheres with the same size, each droplet touches 12 other droplets. The number of contact points will proportional to m rather than m−1. In a heterodisperse system, one droplet may even touch 12 other droplets. This can be taken into account by rewriting equation (11.17) as

$$\frac{dm}{dt} = kn_o - pKm \,, \tag{11.22}$$

where p has a value between 1 and 6. On integrating equation (11.22), one obtains

$$m = \frac{kn_o}{pK} + \left(2 - \frac{kn_o}{pK}\right) \exp\left(-pKt\right), \tag{11.23}$$

which replaces equation (11.18) for concentrated emulsions. This means that for concentrated emulsions the rate of coalescence increases with droplet concentration in a manner dependent on the droplet size distribution, the degree of packing, and on the size of aggregates.

(ii) In a very dilute emulsion, kn_o/K can be much smaller than unity if coalescence occurs very rapidly. After flocculation has proceeded for sufficient time, such that $Kt \gg 1$, the second term on the right-hand side of equation (11.19) may be neglected relative to the first term. This equation reduces to Smoluchowski's equation, i.e. the rate of flocculation is independent on any coalescence.

(iii) If the degree of coalescence is very small, the exponential term in equation (11.19) may be expanded in a power series, retaining the first two terms only when $K \ll 1$, such that the equation reduces to

$$n = n_o \left[1 - \frac{Kt}{(1 + kn_o t)} + \frac{Kt}{(1 + kn_o t)^2} \right]. \tag{11.24}$$

Equation (11.24) predicts only a very small decrease in droplet number concentration with time, as expected.

(iv) When flocculation has proceeded for a sufficiently long time, Kt may be much greater than unity. In this case the exponential term may be neglected, and then $kn_o/t \gg 1$ (in the denominator). Then,

$$n = \frac{n_o}{Kt} + \frac{1}{kt}. \tag{11.25}$$

With large n_o, the first term on the right hand side of equation (11.25) predominates, and hence $1/kt$ can be neglected.

Davies and Rideal [8] discussed the problem of coalescence, incorporating an energy barrier term into the Smoluchowski equation, in order to account for the slow coalescence for emulsions stabilised by sodium oleate. The Smoluchowsi equation may be written in terms of the mean volume V of emulsion droplets:

$$V = \frac{\varphi}{n_o} + 4\pi DR\varphi t, \tag{11.26}$$

where φ is the volume fraction of the dispersed phase, D is the diffusion coefficient of the droplets and R is the collision radius. D can be calculated using the Stokes–Einstein equation:

$$D = \frac{kT}{6\pi\eta R}, \tag{11.27}$$

where k is the Boltzmann constant, T is the absolute temperature, η is the viscosity of the medium and R is the droplet radius.

Equation (11.26) predicts that the mean volume of the droplets should be doubled in about 43 s, whereas experiments show that in the presence of sodium oleate this takes about 50 days. To account for this, an energy barrier (ΔG_{coal}) was introduced into equation (11.26):

$$V = V_o + 4\pi DR\varphi t \exp\left(-\frac{\Delta G_{coal}}{kT}\right). \tag{11.28}$$

Substituting for D from equation (11.27) and differentiating with respect to t, the rate of coalescence for an O/W emulsion is given by

$$\frac{dV}{dt} = \frac{4\varphi kT}{3\eta_w} \exp\left(-\frac{\Delta G_{coal}}{kT}\right) = C_1 \exp\left(-\frac{\Delta G_{coal}}{kT}\right), \tag{11.29}$$

where η_w is the viscosity of the continuous phase (water for O/W emulsion) and C_1 is the collision factor defined by equation (11.29).

For a W/O emulsion, the corresponding relation would be

$$\frac{dV}{dt} = \frac{4(1-\varphi)kT}{3\eta_o} \exp\left(-\frac{\Delta G_{coal}}{kT}\right) = C_2 \exp\left(-\frac{\Delta G_{coal}}{kT}\right),\qquad(11.30)$$

where η_o is the viscosity of the oil continuous phase and C_2 is the corresponding collision factor.

Davies and Rideal considered the energy barrier in terms of the electrical potential ψ_o at the surface of the oil droplets, arising from drops stabilised by ionic surfactants. The energy barrier preventing coalescence is proportional to ψ_o^2, according to the DLVO theory [2, 3], as described above:

$$\Delta G_{coal} = B\psi_o^2,\qquad(11.31)$$

where B is a constant that depends on the radius of curvature of the droplets. When two approaching droplets tend to flatten in the region of contact in a lamella, the radius of curvature to be used for emulsion droplets may be considerably different from the actual droplet radius. However the degree of flattening is negligible for small emulsion droplets ($<1\,\mu$m in diameter). If there is specific adsorption of counter ions, the electric potential to be used in evaluating the electrical double repulsion will be less than ψ_o. In this case one has to use the Stern potential ψ_d at the plane of specifically adsorbed ions (see Chapter 3), i.e.

$$\Delta G_{coal} = B\psi_d^2.\qquad(11.32)$$

11.5 Reduction of coalescence

11.5.1 Use of mixed surfactant films

It has long been known that mixed surfactants can have a synergistic effect on emulsion stability, with respect to coalescence rates. For example, Schulman and Cockbain [9] found that the stability of Nujol/water emulsions increases markedly with the addition of cetyl alcohol or cholesterol to an emulsion prepared using sodium cetyl sulphate. The enhanced stability was assumed to be associated with the formation of a densely packed interfacial layer. The maximum effect is obtained when using a water-soluble surfactant (cetyl sulphate) and an oil-soluble surfactant (cetyl alcohol), sometimes referred to as co-surfactant, are used in combination. Suitable combinations lead to enhanced stability, compared to the individual components. These mixed surfactant films also produce a low interfacial tension, in the region of 0.1 mN m^{-1} or lower. This reduction in interfacial tension may be due to the cooperative adsorption of the two surfactant molecules, as predicted by the Gibbs adsorption equation for multi-component systems,

For a multi-component system i, each with an adsorption Γ_i (mol m^{-2}, referred to as the surface excess), the reduction in γ, i.e dγ, is given by the following expression:

$$d\gamma = -\sum \Gamma_i \, d\mu_i = -\sum \Gamma_i RT \, d\ln C_i \,, \tag{11.33}$$

where μ_i is the chemical potential of component i, R is the gas constant, T is the absolute temperature and C_i is the concentration (mol dm^{-3}) of each surfactant component.

The reason for the lowering of γ when using two surfactant molecules can be understood under consideration of the Gibbs adsorption equation for multi-component systems [9]. For two components s_a (surfactant) and c_o (cosurfactant), equation (11.33) becomes

$$d\gamma = -\Gamma_{sa} RT \, d\ln C_{sa} - \Gamma_{co} RT \, d\ln C_{co} \,. \tag{11.34}$$

Integration of equation (11.34) gives

$$\gamma = \gamma_0 - \int_0^{C_{sa}} \Gamma_{sa} RT \, d\ln C_{sa} - \int_0^{C_{co}} \Gamma_{co} RT \, d\ln C_{co} \,, \tag{11.35}$$

which clearly shows that γ_0 is lowered by two terms, both surfactant and cosurfactant.

The two surfactant molecules should adsorb simultaneously, and they should not interact with each other; otherwise they lower their respective activities. Thus, the surfactant and cosurfactant molecules should vary in nature, one predominantly water soluble (such as an anionic surfactant) and one predominantly oil soluble (such as a long chain alcohol).

Several mechanisms have been suggested to account for the enhanced stability produced by using mixed surfactant films, and these are summarised below.

(i) Interfacial tension and Gibbs elasticity

The synergistic effect of surfactant mixtures can be accounted for by the enhanced lowering of interfacial tension of the mixture when compared with individual components. For example, addition of cetyl alcohol to an O/W emulsion stabilised by cetyl trimethyl ammonium bromide results in lower interfacial tension, and a shift of the critical micelle concentration (CMC) results in lower values, probably due to the increased packing of the molecules at the O/W interface [1]. Another effect of using surfactant mixtures is due to the enhanced Gibbs dilational elasticity, ε:

$$\varepsilon = \frac{d\gamma}{d\ln A} \,, \tag{11.36}$$

where dγ is the change in interfacial tension resulting from increase of the interfacial area dA on expanding the interface.

Prins and van den Tempel [9] showed that the surfactant mixture sodium laurate plus lauric acid gives a very high Gibbs elasticity (of the order of 10^3 mN m^{-1}) when compared with that of sodium laurate alone. In the presence of laurate ions, lauric acid has an extremely high surface activity. At half coverage, the interface contains $1.3\,\mathrm{mol\,dm^{-3}}$ laurate ions and $4.8 \cdot 10^{-7}\,\mathrm{mol\,dm^{-3}}$ lauric acid. Thus, under these conditions, the minor constituent can contribute more to the Gibbs elasticity than the major constituent. Similar results were obtained by Prins et al. [10], who showed that ε increases markedly in the presence of lauryl alcohol for O/W emulsions stabilised by sodium lauryl sulphate. A correlation between film elasticity and coalescence rate has been observed for O/W emulsions stabilised with proteins [11].

(ii) Interfacial viscosity

It has long been assumed that high interfacial viscosity could account for the stability of liquid films. This must play a role under dynamic conditions, i.e. when two droplets approach each other. Under static conditions, the interfacial viscosity does not play a direct role. However, a high interfacial viscosity is often accompanied by a high interfacial elasticity, and this may be an indirect contribution to the increased stability of the emulsion. Prins and van den Tempel [9] argued against there being any role played by the interfacial viscosity, due to two main observations, namely the small changes in film stability with change in temperature (which should have a significant effect on the interfacial viscosity) and the sudden decrease of the interfacial viscosity with a slight increase in the concentration of the major component. Thus, Prins and van den Tempel [8] attributed the enhanced emulsion stability resulting from the presence of a minor component to be solely due to an increase in interfacial elasticity.

(iii) Hindrance to diffusion

Another possible explanation of the enhanced stability in the presence of mixed surfactants could be connected to the hindered diffusion of the surfactant molecules in the condensed film. This would imply that the desorption of surfactant molecules is hindered on the approach of two emulsion droplets, and hence thinning of the film is prevented.

(iv) Liquid crystalline phase formation

Friberg and co-workers [12] attributed the enhanced stability of emulsions formed with mixtures of surfactants to the formation of three-dimensional structures, namely, liquid crystals. These structures can form, for example, in a three-component system of surfactant, alcohol and water, as illustrated in Fig. 11.10. The lamellar liquid crystalline phase, denoted by N (neat phase), in the phase diagram is particularly important for stabilising the emulsion against coalescence. In this case the liquid crystals "wrap" around the droplets in several layers, as will be illustrated below. These multilayers form a barrier against coalescence as will be discussed below. Friberg et al. [12]

have given an explanation in terms of the reduced attractive potential energy between two emulsion droplets, each surrounded by a layer of liquid crystalline phase. They have also considered changes in the hydrodynamic interactions in the interdroplet region; this affects the aggregation kinetics.

Friberg et al. [12] have calculated the effect on the van der Waals attraction of the presence of a liquid crystalline phase surrounding the droplets. A schematic representation of the flocculation and coalescence of droplets with and without a liquid crystalline layer is shown in Fig. 11.11.

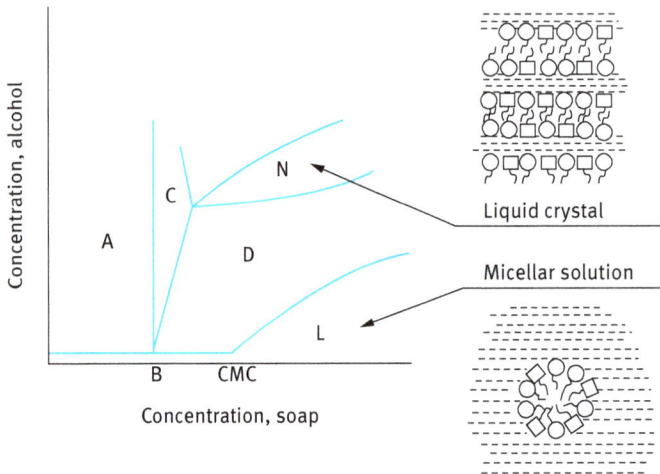

Fig. 11.10: Phase diagram of surfactant-alcohol-water system.

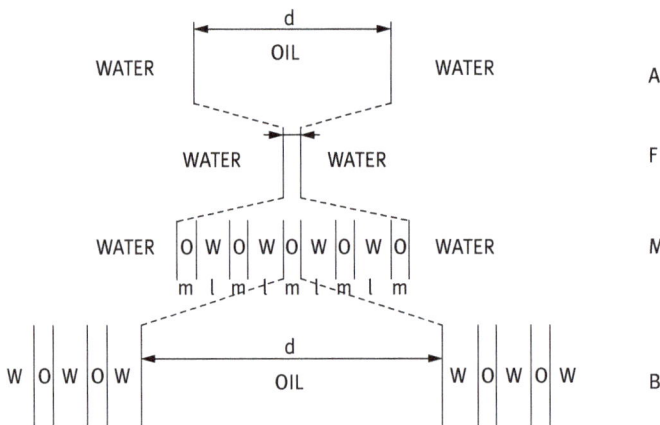

Fig. 11.11: Schematic representation of flocculation and coalescence in the presence and absence of liquid crystalline phases.

The upper part of Fig. 11.11 (A–F) represents the flocculation process when the emulsifier is adsorbed as a mono-molecular layer. The distance d between the water droplets decreases to a distance m, at which film ruptures and the droplets coalesce; m is chosen to correspond to the thickness of the hydrophilic layers in the liquid crystalline phase. This simplifies the calculations and facilitates comparison with the case in which the liquid crystalline layer is adsorbed around the droplets.

The flocculation process for the case of droplets covered with liquid crystalline layers is illustrated in the lower part of Fig. 11.11 (B–M). The oil layer between the droplets thins to thickness m. The coalescence process which follows involves the removal of successive layers between the droplets until a thickness of one layer is reached (F); the final coalescence step occurs in a similar manner to the case for a monomolecular layer of adsorbed surfactant.

For case A, the van der Waals attraction is given by the expression

$$G_A = -\frac{A}{12\pi d^2},$$
(11.37)

where A is the effective Hamaker constant, and

$$A = (A_{11}^{1/2} - A_{22}^{1/2})^2,$$
(11.38)

where A_{11} and A_{22} are the Hamaker constants of the two phases.

For case B, G_B can be obtained from the algebraic summation of this expression for the aqueous layer on each side of the central layer. The ratio B_B/G_A is then given by

$$
\begin{aligned}
\frac{G_B}{G_A} = d^2 \Bigg\{ &\sum_{p=0}^{n}\sum_{q=0}^{n}[d + (p+q)(l+m)]^{-2} \\
&+ \sum_{p=0}^{n-1}\sum_{q=0}^{n-1}[d + 2l + (p+q)(l+m)]^{-2} \\
&- \sum_{p=0}^{n}\sum_{q=0}^{n-1}[2(d+l) + (p+q)(l+m)]^{-2} \Bigg\},
\end{aligned}
$$
(11.39)

where l and m are the thicknesses of the water and oil layers, n is the number of water layers (which is equal to the number of oil layers) and p and q are integers.

The free energy change associated with coalescence (i.e. M → F) is calculated from the variation of the van der Waals interaction across the droplet walls. This treatment reflects the energy change associated with the layers squeezed out from the inter-droplet region. The problem is circumvented by assuming that these displaced layers adhere to the enlarged droplets, so that their free energy is not significantly changed in the process. In this manner the ratio of the interaction energies in the states F and M is obtained from summation of the van der Waals interactions from the individual layers on the water parts, i.e.

$$\frac{G_M}{G_F} = m^2 \Bigg\{ \sum_{p=0}^{n}[(m+p)(m+1)]^{-2} - \sum_{p=0}^{n}[(p+1)(m+1)]^{-2} \Bigg\}.$$
(11.40)

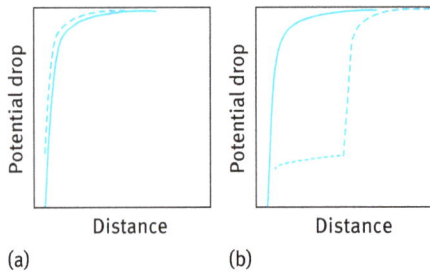

Fig. 11.12: van der Waals potential energy–distance curves for flocculation (a) and coalescence (b) in the presence (– – –) and absence (———) of liquid crystalline phases.

To illustrate the relative importance of the van de Waals attraction energy, calculations were made using the above expressions, for the case of flocculation (i.e., A and B) and for the case of coalescence (i.e. F and M). The results are given in Fig. 11.12 (a) and (b), respectively.

Fig. 11.12 (a) shows that the influence of liquid crystalline layers around the droplets on the flocculation process is insignificant. In contrast, the effect on the free-energy change is quite significant. For example with nine layers on each droplet and a layer thickness of reasonable magnitude ($l = m = 5$ nm), the total van der Waals interaction is reduced to only 10 % of its original value for the coalescence process in the case of the layer structure. This is to be compared with 98 % in the case of two droplets at the same distance, but separated by the oil phase instead of the liquid crystalline phase. Even more important are the extremely small changes in attraction energy after removal of the first layers. The first layers give a drop corresponding to 1.5 % of the total van deer Waals interaction energy (A to F). The last layer, before the state F is reached, corresponds to 78 % of the total van deer Waals energy. It seems, therefore, that the presence of a liquid crystalline phase has a pronounced influence on the distance dependence of the van deer Waals energy, leading to a drastic reduction in the force of attraction between the emulsion droplets.

(v) Use of polymeric surfactants

The most convenient polymeric surfactants are those of the block and graft copolymer type. A block copolymer is a linear arrangement of blocks of variable monomer composition. The nomenclature for a diblock is poly-A-block-poly-B, and for a tailback is poly-A-block-poly-B-poly-A. One of the most widely used tailback polymeric surfactants are the "Plutonic" (BASF, Germany) which consists of two poly-A blocks of poly (ethylene oxide) (PEO) and one block of poly (propylene oxide) (PPO). Several chain lengths of PEO and PPO are available.

The above polymeric tailbacks can be applied as emulsifiers, whereby the assumption is made that the hydrophobic PPO chain resides at the hydrophobic surface, leaving the two PEO chains dangling in aqueous solution and hence providing satiric repulsion, and this reduces or eliminates the coalescence of emulsions.

A graft copolymer based on polysaccharides on insulin, a linear polyfructose chain with a glucose end, has been developed for stabilisation of emulsions [13]. This molecule is used to prepare a series of graft copolymers by random grafting of alkyl chains (using alky isocyanate) on the inulin backbone. The first molecule of this series is INUTEC®SP1, which is obtained by random grafting of C_{12} alkyl chains. It has an average molecular weight of ~5000 Dalton. The molecule is schematically illustrated in Fig. 11.13, which shows the hydrophilic polyfructose chain (backbone) and the randomly attached alky chains.

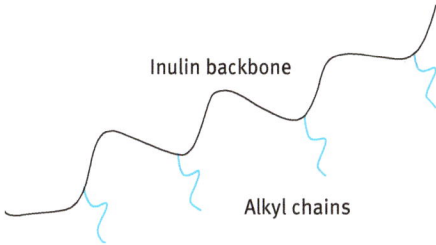

Inulin backbone

Alkyl chains

Fig. 11.13: Schematic representation of INUTEC®SP1 polymeric surfactant.

The main advantages of INUTEC®SP1 as a stabilizer for emulsions are: (i) strong adsorption to the droplet by multi-point attachment with several alky chains. This ensures lack of desorption and displacement of the molecule from the interface; (ii) strong hydration of the linear polyfructose chains both in water and in the presence of high electrolyte concentrations and high temperatures. This ensures effective steric stabilization.

Emulsions of Isopar M/water and cyclomethicone/water were prepared using INUTEC®SP1. 50/50 (v/v) O/W emulsions were prepared and the emulsifier concentration was varied from 0.25 to 2 (w/v) % based on the oil phase. 0.5 (w/v) % emulsifier was sufficient for stabilization of these 50/50 (v/v) emulsions [13]. The emulsions were stored at room temperature and 50 °C, and optical micrographs were taken at intervals of time (for a year) in order to check the stability. Emulsions prepared in water were very stable, showing no change in droplet size distribution over a period of more than a year, and this indicated absence of coalescence. Any weak flocculation that occurred was reversible, and the emulsion could be redispersed by gentle shaking. Fig. 11.14 shows an optical micrograph for a dilute 50/50 (v/v) emulsion that was stored for 1.5 and 14 weeks at 50 °C. No change in droplet size was observed after storage for more than 1 year at 50 °C, indicating the absence of coalescence. The same result was obtained when using different oils. Emulsions were also stable against coalescence in the presence of high electrolyte concentrations (up to 4 mol dm^{-3} or ~25 % NaCl). This stability in high electrolyte concentrations is not observed with polymeric surfactants based on polethylene oxide. The high stability observed using INUTEC®SP1 is related to its strong hydration both in water and in electrolyte solutions. The hydration of inulin (the backbone of HMI) could be assessed using cloud point measurements.

(a)   (b)

Fig. 11.14: Optical micrographs of O/W emulsions stabilized with INUTEC®SP1 stored at 50 °C for 1.5 weeks (a) and 14 weeks (b).

A comparison was also made with PEO with two molecular weights, namely 4 000 and 20 000.

Evidence for the high stability of the liquid film between emulsion droplets when using INUTEC®SP1 was obtained by Exerowa et al. [14], using disjoining pressure measurements. This is illustrated in Fig. 11.15 which shows a plot of disjoining pressure vs separation distance between two emulsion droplets at various electrolyte concentra-

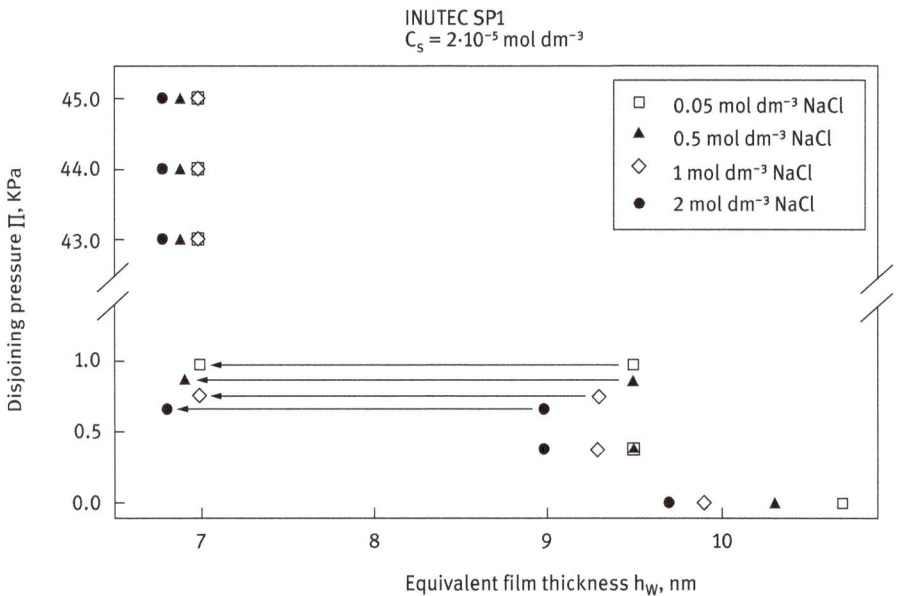

Fig. 11.15: Variation of disjoining pressure with equivalent film thickness at various NaCl concentrations.

tions. The results show that by increasing the capillary pressure a stable Newton black film (NBF) is obtained at a film thickness of ~ 7 nm. The lack of rupture of the film at the highest pressure applied of $4.5 \cdot 10^4$ Pa indicate the high stability of the film in water and in high electrolyte concentrations (up to 2.0 mol dm^{-3} NaCl).

The lack of rupture of the NBF up to the highest pressure applied, namely $4.5 \cdot 10^4$ Pa clearly indicates the high stability of the liquid film in the presence of high NaCl concentrations (up to 2 mol dm^{-3}). This result is consistent with the high emulsion stability obtained at high electrolyte concentrations and high temperature. Emulsions of Isopar M-in-water are very stable under such conditions, and this could be accounted for by the high stability of the NBF. The droplet size of 50 : 50 O/W emulsions prepared using 2 % INUTEC®SP1 is in the region of 1–10 μm. This corresponds to a capillary pressure of ~ $3 \cdot 10^4$ Pa for the 1 μm drops and ~ $3 \cdot 10^3$ Pa for the 10 μm drops. These capillary pressures are lower than those to which the NBF have been subjected to, and this clearly indicates the high stability obtained against coalescence in these emulsions.

References

[1]   Th. F. Tadros and B. Vincent, in: P. Becher (ed.), *Encyclopedia of Emulsion Technology*, Marcel Dekker, New York, 1983.

[2]   B. V. Deryaguin and L. Landau, *Acta Physicochem. USSR* **14** (1941), 633.

[3]   E. J. W. Verwey and J. T. G. Overbeek, *Theory of Stability of Lyophobic Colloids*, Elsevier, Amsterdam, 1948.

[4]   B. V. Deryaguin and R. L. Scherbaker, *Kolloid Zh.* **23** (1961), 33.

[5]   A. Scheludko, *Advances Colloid Interface Sci.* **1** (1967), 391.

[6]   A. Vrij, *Discussion Faraday Soc.* **42** (1966), 23.

[7]   M. van den Tempel, *Rec. Trav. Chim.* **72** (1953), 433, 442.

[8]   J. H. Schulman and E. G. Cockbain, *Transaction Faraday Soc.* **36** (1940), 661.

[9]   A. Prince and M. van den Tempel, *Proc. 4th Int. Congr. Surface Activity, vol. II*, Gordon and Breach, Londo, 1967, p. 1119.

[10] A. Prince, C. Arcuri and M. van den Tempel, *J. Colloid and Interface Sci.* **24** (1967), 811.

[11] B. Biswas and D. A. Haydon, *Proc. Roy. Soc.* **A271** (1963), 296; *Proc. Roy. Soc.* **A2** (1963), 317.

[12] S. Friberg, P. O. Jansson and E. Cederberg, *J. Colloid Interface Sci.* **55** (1976), 614.

[13] Th. F. Tadros, A. Vandamme, B. Levecke, K. Booten and C. V. Stevens, *Advances Colloid Interface Sci.* **108–109** (2004), 207.

[14] D. Exerowa, G. Gotchev, T. Kolarev, K. Khristov, B. Levecke and T. Tadros, *Langmuir* **23** (2007), 1711.

12 Phase inversion and its prevention

12.1 Introduction

Phase inversion is the process whereby the internal and external phase of an emulsion suddenly invert, e.g., O/W to W/O or vice versa [1, 2]. Catastrophic inversion is induced by increasing the volume fraction of the disperse phase. This type of inversion is not reversible [2]; the value of the water : oil ratio at the transition when oil is added to water is not the same as that when water is added to oil. The inversion point depends on the intensity of agitation and the rate of liquid addition to the emulsion.

Phase inversion can also be transitional induced by changing facers which affect the HLB of the system, e.g. temperature and/or electrolyte concentration. The average droplet size decreases and the emulsification rate (defined as the time required to achieve a stable droplet size) increases as inversion is approached. Both trends are consistent with O/W interfacial tension reaching a minimum near the inversion point.

12.2 Catastrophic inversion

Catastrophic inversion is illustrated in Fig. 12.1, which shows the variation of viscosity and conductivity with the oil volume fraction φ. As can be seen, inversion occurs at a critical φ, which may be identified with the maximum packing fraction [1]. At φ_{cr}, η suddenly decreases; the inverted W/O emulsion has a much lower volume fraction. κ also decreases sharply at the inversion point since the continuous phase is now oil. Similar trends are observed if water is added to a W/O emulsion, but in this case the conductivity of the emulsion increases sharply at the inversion point. For example, if one starts with a W/O emulsion, then, on increasing the volume fraction of the water phase (the disperse phase), the viscosity of the emulsion increases gradually until a maximum value is obtained, generally ~0.74. When the inversion takes place to an

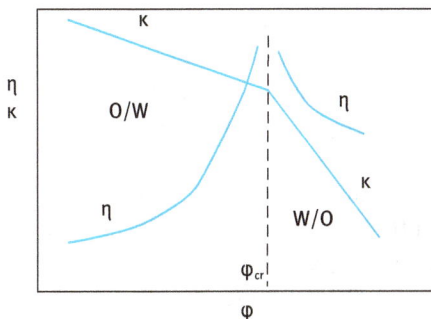

Fig. 12.1: Variation of conductivity and viscosity with volume fraction of oil.

O/W emulsion, the volume fraction of the disperse phase (the oil) will now be ~ 0.26; hence the dramatic decrease in viscosity.

In the early theories of phase inversion, it was postulated that the inversion takes place as a result of the difficulty in packing the emulsion droplets above a certain volume fraction. For example, according to Ostwald [3] an assembly of spheres of equal radii should occupy 74 % of the total volume. Thus, at phase volume $\varphi > 0.74$, the emulsion droplets have to be packed more densely than is possible.

This means that any attempt to increase the phase volume beyond that point should result in distortion, breaking or inversion. However, several investigations showed the invalidity of this argument, inversion being found to take place at phase volumes much greater or smaller than this critical value. For example, Shinoda and Saito [4] showed that inversion of olive oil/water emulsions takes place at $\varphi = 0.25$. Moreover, Sherman [5] showed that the volume fraction at which inversion takes place depends to a large extent on the nature of the emulsifier. It should be mentioned that Ostwald's theory [3] applies only to the packing of rigid, non-deformable spheres of equal size. Emulsion droplets are nether resistant to deformation, nor are they, in general, of equal size. The wide distribution of droplet size makes it possible to achieve a higher internal phase volume fraction by virtue of the fact that the smaller droplets can be fitted into the interstices between the larger ones. If one adds to this the possibility that the droplets may be deformed into polyhedra, even denser packing is possible. This is the principle of preparing high internal phase emulsions (HIPE) reaching $\varphi > 0.95$.

A useful index to characterise phase inversion is to measure the emulsion inversion point (EIP). The EIP is related to the inversion of W/O emulsions to O/W emulsions at constant temperature [6]. An aqueous phase is added (incrementally) to a finite amount of oil which contains a known amount of surfactant. The mixture is agitated by a turbine blender for 15 s on each addition and the emulsion type is determined. The EIP is simply the ratio of the volume of the aqueous phase at the inversion point to the volume of the oil phase. A plot of EIP vs HLB is made, and the results show that the EIP decreases with increase of the HLB number until a minimum is observed. The value of the HLB at the minimum is the required HLB of the oil to produce a stable emulsion. However, the exact position of the EIP minimum can be affected by the agitation conditions.

Several findings of the EIP experiments have been found. (i) At the EIP minimum, the inversion from W/O to O/W occurs and produces emulsions with very small drops. (ii) The EIP increases with increase of lipophilic surfactant, whereas the EIP decreases with increase of the concentration of hydrophilic surfactant. (iii) In a series of alkanes, the higher the EIP the lower the required HLB. (iv) Highest viscosity and lowest interfacial tension occur at the EIP. (v) For aromatic hydrocarbons, with increasing methyl group substitutions the EIP and the required HLB decrease. (vi) The EIP shows changes in the required HLB of an oil brought about by addition of additives, e.g., alcohols and poly(ethylene glycol).

When catastrophic inversion is brought about by addition of the aqueous phase to the oil phase (high HLB), two drop types can be present before phase inversion: (i) unstable water drops containing surfactant micelles, in a continuous oil phase (i.e., W_m/O), and (ii) stable oil drops within water drops, in a continuous oil phase (i.e., $O/W_m/O$). When catastrophic inversion is brought about by adding of the oil phase to the water phase (low HLB), two drop types can be present before phase inversion: (i) unstable oil drops containing surfactant micelles, in a continuous aqueous phase (i.e., O_m/W), and (ii) stable water drops within oil drops, in a continuous aqueous phase (i.e., $W/O_m/W$).

After catastrophic inversion has taken place the resulting emulsion consists of stable oil drops in a continuous water phase containing surfactant micelles (i.e., O/W_m) when the initial continuous phase is oil. When the initial continuous phase is aqueous, the resulting emulsion consists of stable water drops in a continuous oil phase containing surfactant micelles (i.e., W/O_m).

Ostwald [3] first modelled catastrophic inversions as being caused by the complete coalescence of the dispersed phase at the close packed condition, whereas Marzall [7] has shown that catastrophic inversion can occur over a wide range of water : oil ratio. This may be due to the formation of double emulsion drops ($O/W_m/O$) boosting the actual volume of the dispersed phase.

The dynamic factors affecting catastrophic inversion have been concerned with the movement of inversion boundaries, with either changes in the system composition or changes in the system's dynamics, such as the effect of agitation conditions. For systems that do not contain stabilising surfactant, inversion hysteresis has been shown to occur. As the viscosity of the oil phase increases, the more likely it is that the oil becomes the dispersed phase. Inversion is shifted to a higher dispersed phase fraction as the stirrer speed increases.

12.3 Transitional inversion

Transitional inversion is caused by change of the system HLB at constant temperature using surfactant mixtures. This is illustrated in Fig. 12.2, which shows the change in the droplet Sauter diameter, d_{32} (volume/area mean diameter) and rate constant (min^{-1}) as a function of the HLB of (non-ionic) surfactant mixtures [2].

It can be seen from Fig. 12.2 that the average droplet diameter decreases and the emulsification rate (defined as the time required to achieve a stable droplet size) increases as inversion is approached. The results are consistent with the oil/water interfacial tension passing through a minimum within the HLB range where the three-phase forms. It was also noted that the emulsion formed by transitional inversion are finer, and that they require less energy than those made by direct emulsification.

Several other conditions can cause transitional phase inversion, such as addition of electrolyte and/or increase of temperature, in particular for emulsions based on

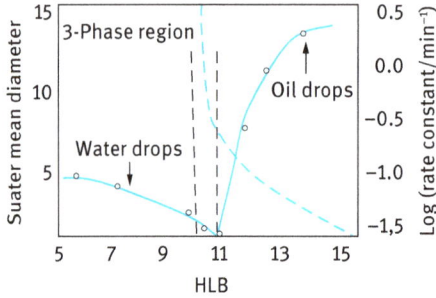

Fig. 12.2: Emulsion drop diameters (circles) and rate constant for attending steady size (squares) as a function of surfactant HLB in cyclohexane/0.067 mol dm^{-3} KCl containing nonylphenol ethoxylates at 25 °C.

non-ionic surfactants of the ethoxylate. In order to understand the process of phase inversion, one must consider the surfactant affinity to the oil and water phases, as described by the Winsor R_o ratio [8], which is the ratio of the intermolecular attraction of oil molecules (O) and lipophilic portion of surfactant (L), C_{LO}, to that of water (W) and hydrophilic portion (H), C_{HW}:

$$R_o = \frac{C_{LO}}{C_{HW}} . \tag{12.1}$$

Several interaction parameters may be identified at the oil and water sides of the interface. One can identify at least nine interaction parameters as was schematically represented in Fig. 7.10 of Chapter 7. This figure is reproduced here as Fig. 12.3 for the sake of clarification.

C_{LL}, C_{OO}, C_{LO} (at oil side)

C_{HH}, C_{WW}, C_{HW} (at water side)

C_{LW}, C_{HO}, C_{LH} (at the interface)

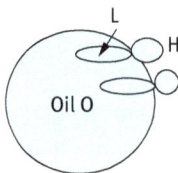

Fig. 12.3: Various interaction parameters at the oil and water phases.

In Fig. 12.3, C_{LL}, C_{OO} and C_{LO} refer to the interaction energies between the two lipophilic parts of the surfactant molecule, the interaction energy between two oil molecules and the interaction energy between the lipophilic chain and oil, respectively. C_{HH}, C_{WW} and C_{HW} refer to the interaction energies between the two hydrophilic parts of the surfactant molecule, the interaction energy between two water molecules and the interaction energy between the hydrophilic chain and water, respectively. C_{LW}, C_{HO} and C_{LH} refer to the interaction energies at the interface between the lipophilic part of the surfactant molecule and water, the interaction energy between the hydrophilic part and oil and the interaction energy between the lipophilic and hydrophilic parts, respectively.

The three cases $R_o < 1$, $R_o > 1$ and $R_o = 1$ correspond to type I (O/W), type II (W/O) and type III (flat interface) phase behaviour, respectively. For example, for $R_o < 1$, an increase of temperature results in the reduction of the hydration of the hydrophilic part of the surfactant molecule, and the emulsion changes from Winsor I to Winsor III to Winsor II, and this causes phase inversion from O/W to W/O emulsion. This inversion occurs at a particular temperature, referred to as the phase inversion temperature (PIT), as will be discussed below.

In Winsor's type I systems ($R_o < 1$), the affinity of the surfactant for the water phase exceeds its affinity to the oil phase. Thus, the interface will be convex towards water, and the non-ionic surfactant-oil-water (n-SOW) system can be one or two phases. A system in the two-phase system will split into an oil phase containing dissolved surfactant monomers at the CMC_o (critical micelle concentration in the oil phase) and an aqueous microemulsion-water phase containing solubilised oil in normal surfactant micelles.

In Winsor's type II systems ($R_o > 1$), the affinity of the surfactant for the oil phase exceeds its affinity to the water phase. Thus, the interface will be convex towards oil, and the non-ionic surfactant-oil-water (n-SOW) system can be one or two phases. A system in the two-phase system will split into a water phase containing dissolved surfactant monomers at the CMC_w (critical micelle concentration in the water phase) and an oleic microemulsion phase containing solubilised water in inverse surfactant micelles.

In Winsor's type III system ($R_o = 1$), the surfactants affinity for the oil and water phases is balanced. The interface will be flat, and the n-SOW system can have one, two or three phases, depending on its composition. In the multi-phase region, the system can be: (i) two-phase, a water phase and an oleic microemulsion; (ii) two-phase, an oil phase and an aqueous microemulsion; (iii) three-phase, a water phase containing surfactant monomers at CMC_w, an oil phase containing surfactant monomers at CMC_o and a "surfactant phase". The latter phase may have a bicontinuous structure, being composed of cosolubilised oil and water separated from each other by a layer of surfactant. The "surfactant phase" is sometimes called the middle phase, because its intermediate density causes it to appear between the oil and water phases in a phase-separated type III n-SOW system.

12.4 The phase inversion temperature (PIT)

One way to alter the affinity in an n-SOW system is by changing the temperature which changes the surfactant's affinity to the two phases. At high temperature the non-ionic surfactant becomes soluble in the oil phase, whereas at low temperature it becomes more soluble in the water phase. Thus, at a constant surfactant concentration, the phase behaviour will change with temperature [9]. With an increase in temperature the surfactant's affinity to the oil phase increases and the system changes from Winsor I to Winsor III, and finally to Winsor II, i.e. the emulsion will invert from an O/W to W/O system at a particular temperature, referred to as the phase inversion temperature (PIT). A schematic representation of the change in phase behaviour in n-SOW system is shown in Fig. 12.4. At low temperature, over the Winsor I region, O/W emulsions can be easily formed and are quite stable, as shown in Fig. 12.4 (a). On raising the temperature (as shown by the arrow in Fig. 12.4) the O/W emulsion stability decreases, and the macroemulsion finally resolves when the system reaches the Winsor III state (represented in Fig. 12.4 (c)). Within this region, both O/W and W/O emulsions are unstable, with the minimum stability in the balanced region. At higher temperature, over the Winsor II region, W/O emulsions become stable, as represented in Fig. 12.4 (e). This behaviour is always observed in non-ionic systems if the surfactant concentration is above the CMC and the volume fractions of the components are not extreme. The macro-emulsion stability is essentially symmetrical with respect to the balanced point, just as the phase behaviour is. At positive spontaneous curvature, O/W emulsions are stable, while at negative spontaneous curvature, W/O emulsions are stable.

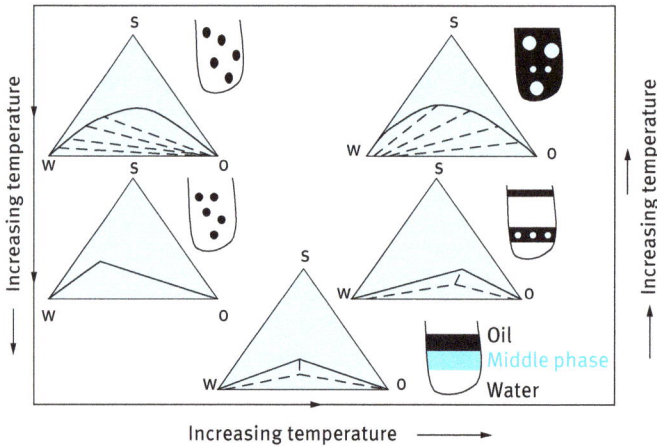

Fig. 12.4: Effect of increasing temperature on the phase behaviour of nSOW system. The PIT concept.

Fig. 12.5 represents the most clear-cut image of the macro-emulsion inversion as a function of temperature [9]. Equal volumes of oil and water are emulsified at various temperatures. Five hours after preparation, the macro-emulsions completely cream or sediment, depending on the temperature. Below the balanced temperature, a stable O/W cream layer is formed, which does not show any visible coalescence. Similarly, above the balanced point a stable sediment of a W/O emulsion is formed. Close to

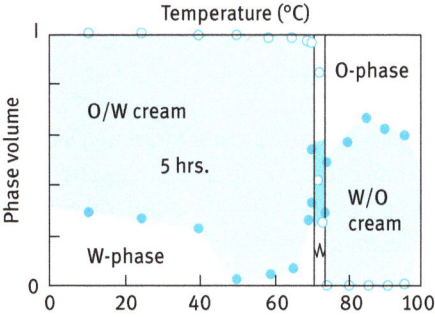

Fig. 12.5: Macroemulsion stability diagram of cyclohexane-water-polyoxyethylene (9.7) nonyl-phenol-ether system.

the balanced point, the narrow temperature range between 66 and 68 °C, where the three-phase equilibrium is observed, neither O/W or W/O macroemulsions are stable.

Within the Winsor III region, the stability of the macroemulsions is very temperature sensitive. Although exactly in the balanced state, the macro-emulsions are very unstable and break within minutes; the system becomes stable only several tens of degrees away from the balanced point, while still being in the Winsor III region. The macro-emulsion stability pattern is not completely symmetric. W/O emulsions reach maximum stability at ~ 20 °C above the balanced point, after which the stability begins to decrease. On the other hand, there is no similar maximum stability for the O/W emulsion stability at very low temperatures.

The macro-emulsion phase behaviour can be "tuned" not only by changes in temperature, but also by addition of "co-solvents", co-surfactants or electrolytes [9]. For example, the balanced point of n-C_8H_{18}–$C_{10}E_5$-water system is at ~ 45 °C, while that of the n-C_8H_{18}–$C_{10}E_5$-10 % NaCl is at ~ 28 °C. The changes in the macro-emulsion phase behaviour induced by additives leads to a similar shift in the macro-emulsion stability profile. Thus, when 10 % of NaCl is added to the system, the new balanced point is now established at 28 °C, and now macroemulsions prepared below 28 °C will have an O/W type. The same effect is found when the balanced location point is controlled by adding co-solvent to oil and water, changing the chain length of the oil, and adding cosurfactants.

Several empirical equations have been proposed to evaluate the location of the balanced point. For example, Salager et al. [10] proposed the following empirical equation for determining the "optimum surfactant formulation" (i.e., the balanced point) for anionic surfactant systems:

$$\ln S - K \cdot ACN - f(A) + \sigma - a_T \Delta T = 0, \tag{12.2}$$

where S is the wt % of NaCl, ACN is the alkane carbon number (a characteristic parameter of the oil phase), ΔT is the temperature deviation from a certain reference (25 °C), f(A) is a function of the alcohol type and concentration, and K, σ and a_T are empirical parameters characterising the surfactant.

A similar empirical equation was proposed for non-ionic surfactants [10]:

$$\alpha - EON + bS - kACN - \varphi(A) + c_T\Delta T = 0. \tag{12.3}$$

EON is the average number of ethylene oxide groups per surfactant molecule, $\varphi(A)$ is another empirical function of the alcohol type and concentration, and α, c_T and k are empirical constants characterising the surfactant. It is clear that the left-hand sides of equations (12.2) and (12.3) are proportional to the mono-layer spontaneous curvature, taken with the opposite sign.

The macro-emulsions invert, as any of the composition parameters is continuously varied in such a way that that the system passes through the optimal composition (balanced state), and the left-hand side of equations (12.2) and (12.3) changes sign. As the spontaneous curvature is varied by changing the mole fraction of one of the surfactants in the mixture, composition of the oil phase, or the mole fraction of the alcohol, the system passes through the Winsor I–Winsor III–Winsor II sequence. This is illustrated in Fig. 12.6, where n-pentanol is added to a 50 : 50 O/W emulsion of kerosene/2 wt % NaCl using an anionic surfactant as emulsifier. It can be seen that the O/W emulsion reaches the balanced point at ~ 5.5 % pentanol, where the emulsion stability (as measured by the lifetime of the emulsion) reaches a minimum (of several

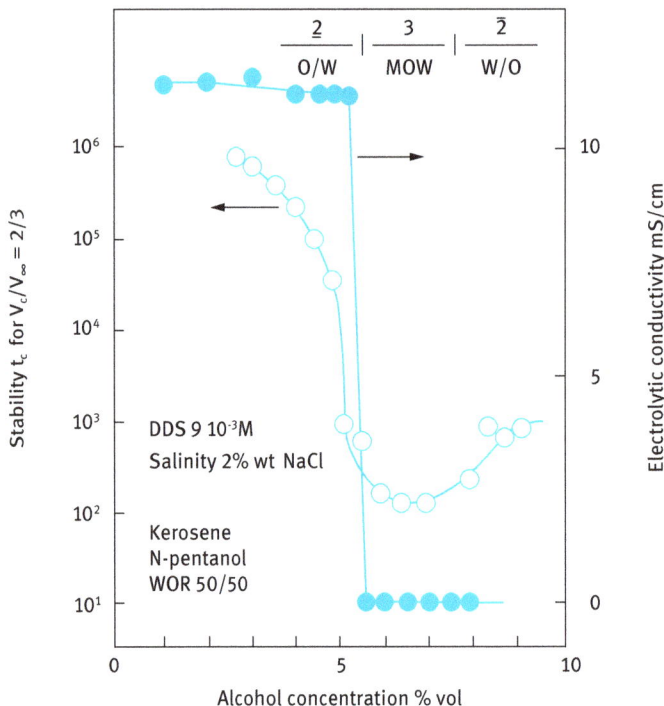

Fig. 12.6: Macro-emulsion inversion caused by addition of n-pentanol.

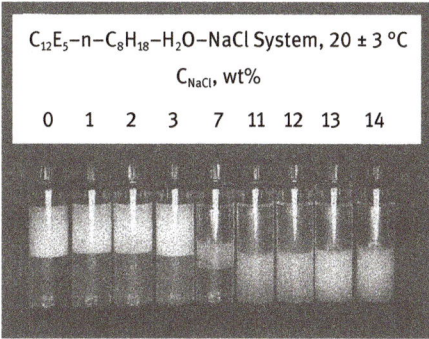

Fig. 12.7: Visual observation of the emulsion type as a function of NaCl concentration.

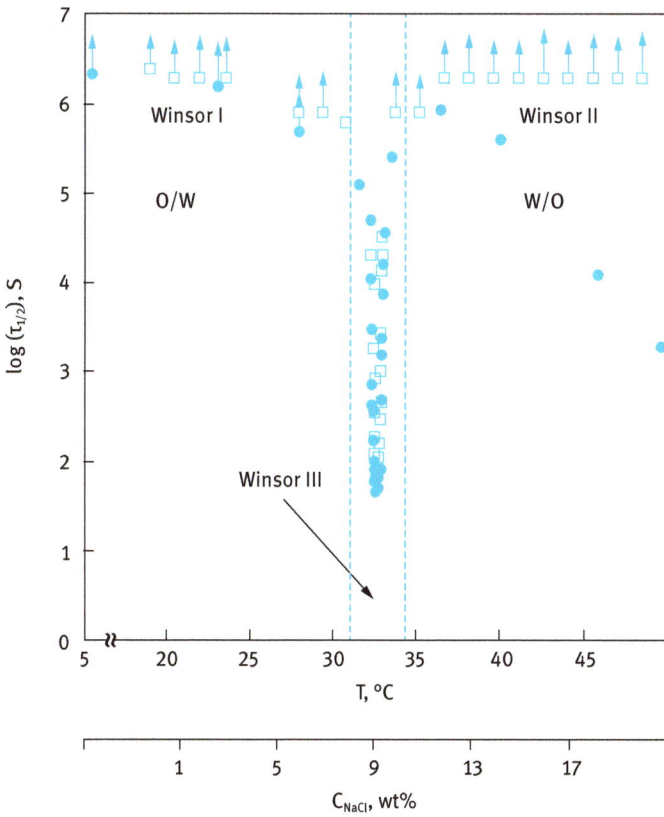

Fig. 12.8: Logarithmic macro-emulsion lifetime (log $\tau_{1/2}$) vs temperature or NaCl concentration.

minutes) at the balanced point. Any addition of pentanol above the balanced point causes inversion of the emulsion, as indicated by the rapid decrease of the conductivity of the emulsion. The lifetime of the emulsion can reach several hours or days, as one moves away from the balanced point.

The overall pattern of macro-emulsion stability as a function of salinity or temperature is illustrated by considering an O/W emulsion of n-octane/water stabilised by a non-ionic surfactant such as $C_{12}E_5$. Fig. 12.7 shows a visual inspection of the emulsion as a function of NaCl concentration at room temperature. At low salinities, the macroemulsion has an O/W type. As the salinity increases, the system changes from very stable O/W to very stable W/O type, with the inversion at the three-phase equilibrium range. O/W emulsions can be distinguished from W/O by the fact that the former forms a cream layer at the top of the container, while the latter forms a milky sediment at the bottom of the container. A schematic representation of the emulsion inversion at increasing temperature or increasing NaCl concentration (at room temperature) is shown in Fig. 12.8, which shows the variation of logarithmic lifetime ($\log \tau_{1/2}$) with increasing temperature or NaCl concentration. Both O/W and W/O emulsions are very

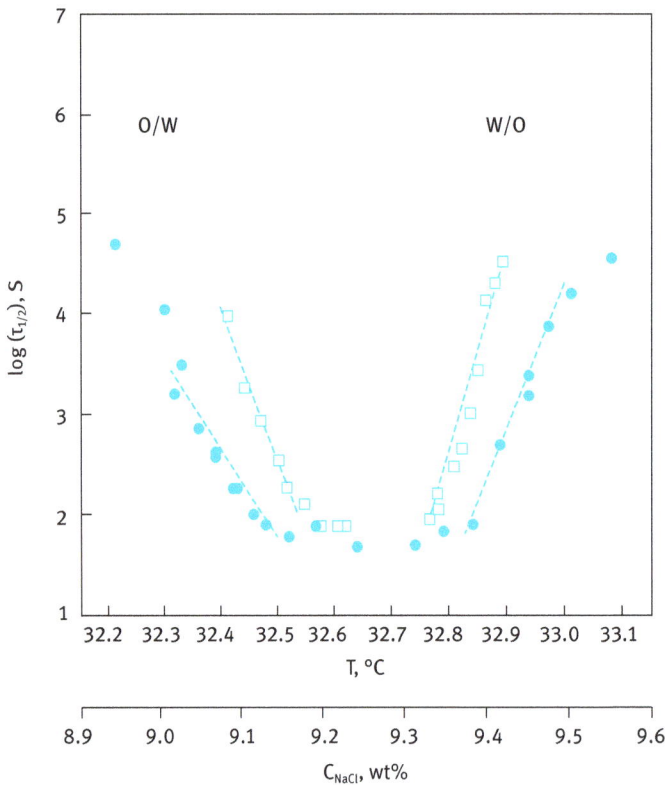

Fig. 12.9: $\log \tau_{1/2}$ vs temperature (upper curve) or NaCl concentration (lower curve).

stable far away from the balanced point. The behaviour is not completely identical when the spontaneous curvature is controlled by temperature. Although at low temperature, the O/W emulsion is very stable, the W/O emulsion stability passes through the maximum and then decreases.

In the Winsor III region, the macroemulsion is extremely temperature- and salinity-sensitive. This is illustrated in Fig. 12.9, which shows that changes by only several tenths of a degree or several tenths of NaCl produce the macroemulsion stability from minutes to days.

References

[1]   Th. F. Tadros and B. Vincent, in: P. Becher (ed.), *Encyclopedia of Emulsion Technology*, Marcel Dekker, New York, 1983.

[2]   B. P. Binks, Emulsions – Recent Advances in Understanding, in: B. P. Binks (ed.), *Modern Aspects of Emulsion Science*, The Royal Society of Chemistry Publication, Cambridge, 1998.

[3]   W. O. Ostwald, *Kolloid Z.* **6** (1910), 103; *Kolloid Z.* **7** (1910), 64.

[4]   K. Shinoda and H. Saito, *J. Colloid Interface Sci.* **30** (1969), 258.

[5]   P. Sherman, J., *Chem. Inc. (London)*, **69** (Suppl. No. 2) (1950), 570.

[6]   B. W. Brooks, H. N. Richmond and M. Zefra, Phase Inversion and Drop Formation in Agitated Liquid-Liquid Dispersions, in: B. P. Binks (ed.), *Modern Aspects of Emulsion Science*, The Royal Society of Chemistry Publication, Cambridge, 1998.

[7]   L. Marzall, in: M. J. Schick (ed.), *Nonionic Surfactants: Physical Chemistry*, Surfactant Science Series 23, Dekker, New York, 1967.

[8]   P. A. Winsor, *Trans. Faraday Soc.* **44** (1948), 376.

[9]   A. S. Kabalnov, Coalescence in Emulsions, in: B. P. Binks (ed.), *Modern Aspects of Emulsion Science*, The Royal Society of Chemistry Publication, Cambridge, 1998.

[10] J. L. Salager, J. Morgan, R. Schechter, W. Wade and E. Vasquez, *Soc. Petrol. Eng. J.* **19** (1979), 107.

13 Characterization of emulsions and assessment of their stability

13.1 Introduction

For full characterization of the properties of emulsions, three main types of investigations are needed:

(i) Fundamental investigation of the system at a molecular level. This requires investigations of the structure of the liquid/liquid interface, namely the structure of the electrical double layer (for charge stabilized emulsions), adsorption of surfactants, polymers and polyelectrolytes and conformation of the adsorbed layers (e.g. the adsorbed layer thickness). It is important to know how each of these parameters changes with the conditions, such as temperature, solvency of the medium for the adsorbed layers and effect of addition of electrolytes.

(ii) Investigation of the sate of emulsion on standing, namely creaming or sedimentation, flocculation rates, flocculation points with sterically stabilised systems, Ostwald ripening and coalescence. All these phenomena require accurate determination of the droplet size distribution as a function of storage time.

(iii) Bulk properties of the suspension, which is particularly important for concentrated emulsions. This requires measurement of the rate of creaming/sedimentation and equilibrium cream/sediment height. More quantitative techniques are based on assessment of the rheological properties of the emulsion (without disturbing the system, i.e. without its dilution and measurement under conditions of low deformation). and how these are affected by long-term storage.

In this chapter I begin with a summary of the methods that can be applied to assess the structure of the solid/liquid interface. This is followed by a more detailed section on assessment of creaming/sedimentation, flocculation, Ostwald ripening and coalescence. For the latter (flocculation, Oswald ripening and coalescence) one needs to obtain information on the droplet size distribution. Several techniques are available for obtaining this information on diluted systems. It is essential to dilute the concentrated emulsion with its own dispersion medium in order not to affect the state of the emulsion during examination. The dispersion medium can be obtained by centrifugation of the emulsion, whereby the supernatant liquid produced at the top or bottom of the centrifuge tube. Care should be taken dilution of the concentrated system with its supernatant liquid (i.e. with minimum shear).

13.2 Assessment of the structure of the liquid/liquid interface

13.2.1 Double layer investigation

The most convenient method for double layer investigation is the electrokinetic and zeta potential measurements [1]. There are essentially two techniques for measurement of the electrophoretic mobility and zeta potential, namely the ultramicroscopic and the velocimetric technique. The latter is the most convenient method, being fast and accurate. This method is suitable for small particles that undergo Brownian motion [1]. The scattered light by small particles will show intensity fluctuation as a result of the Brownian diffusion (Doppler shift). When a light beam passes through a colloidal dispersion, an oscillating dipole movement is induced in the particles, thereby radiating the light. Due to the random position of the particles, the intensity of scattered light, at any instant, appear as random diffraction ("speckle" pattern). As the droplets undergo Brownian motion, the random configuration of the pattern will fluctuate, so that the time taken for an intensity maximum to become a minimum (the coherence time), corresponds approximately to the time required for a particle to move one wavelength λ. Using a photo-multiplier of active area about the diffraction maximum (i.e. one coherent area), this intensity fluctuation can be measured. The analogue output is digitised (using a digital correlator), measuring the photocount (or intensity) correlation function of scattered light. The intensity fluctuation is schematically illustrated in Fig. 13.1.

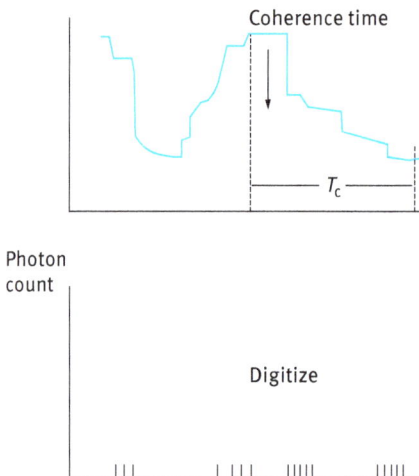

Fig. 13.1: Schematic representation of intensity fluctuation of scattered light.

The photocount correlation function $g^{(2)}(\tau)$ is given by

$$g^{(2)} = B[1 + \gamma^2 g^{(1)}(\tau)]^2 , \qquad (13.1)$$

where τ is the correlation delay time.

The correlator compares $g^{(2)}(\tau)$ for many values of τ. B is the background value to which $g^{(2)}(\tau)$ decays at long delay times. $g^{(1)}(\tau)$ is the normalised correlation function of the scattered electric field and γ is a constant (~ 1).

For mono-dispersed non-interacting droplets,

$$g^{(1)}(\tau) = \exp(-\Gamma\gamma) . \qquad (13.2)$$

Γ is the decay rate or inverse coherence time, which is related to the translational diffusion coefficient D:

$$\Gamma = DK^2 , \qquad (13.3)$$

where K is the scattering vector:

$$K = \left(\frac{4\pi n}{\lambda_0}\right) \sin\left(\frac{\theta}{2}\right) . \qquad (13.4)$$

The droplet radius R can be calculated from D using the Stokes–Einstein equation:

$$D = \frac{kT}{6\pi\eta_0 R} , \qquad (13.5)$$

where η_0 is the viscosity of the medium.

If an electric field is placed at right angles to the incident light and in the plane defined by the incident and observation beam, the line broadening is unaffected, but the centre frequency of the scattered light is shifted to an extent determined by the electrophoretic mobility. The shift is very small compared to the incident frequency (~ 100 Hz for and incident frequency of $\sim 6 \cdot 10^{14}$ Hz), but with a laser source it can be detected by heterodyning (i.e. mixing) the scattered light with the incident beam and detecting the output of the difference frequency. A homodyne method may be applied in which case a modulator to generate an apparent Doppler shift at the modulated frequency is used. To increase the sensitivity of the laser doppler method, the electric fields are much higher than those used in conventional electrophoresis. The Joule heating is minimised by pulsing the electric field in opposite directions. The Brownian motion of the particles also contributes to the Doppler shift, and an approximate correction can be made by subtracting the peak width obtained in the absence of an electric field from the electrophoretic spectrum. A He-Ne Laser is used as the light source, and the output of the laser is split into two coherent beams which are cross-focused in the cell to illuminate the sample. The light scattered by the particle, together with the reference beam, is detected by a photo-multiplier. The output is amplified and analysed to transform the signals to a frequency distribution spectrum. At the intersection of the beam interferences of known spacing are formed.

The magnitude of the Doppler shift Δv is used to calculate the electrophoretic mobility u using the following expression:

$$\Delta v = \left(\frac{2n}{\lambda_0}\right) \sin\left(\frac{\theta}{2}\right) uE, \tag{13.6}$$

where n is the refractive index of the medium, λ_0 is the incident wavelength in vacuum, θ is the scattering angle and E is the field strength.

Several commercial instrument are available for measuring the electrophoretic light scattering:

(i) The Coulter DELSA 440SX (Coulter corporation, USA) is a multi-angle laser Doppler system employing heterodyning and autocorrelation signal processing. Measurements are made at four scattering angles (8, 17, 25 and 34), and the temperature of the cell is controlled by a Peltier device. The instrument reports the electrophoretic mobility, zeta potential, conductivity and particle size distribution.

(ii) Malvern (Malvern Instruments, UK) has two instruments: the ZetaSizer 3000 and ZetaSizer 5000: The ZetaSizer 3000 is a laser Doppler system which uses a crossed beam optical configuration and homodyne detection with photon correlation signal processing. The zeta potential is measured using laser Doppler velocimetry, and the particle size is measured using photon correlation spectroscopy (PCS). The ZetaSizer 5000 uses PCS to measure both (a) movement of the particles in an electric field for zeta potential determination and (b) random diffusion of particles at different measuring angles for size measurement on the same sample. The manufacturer claims that zeta potential for particles in the range 50 nm to 30 µm can be measured. In both instruments, and a Peltier device is used for temperature control.

13.2.2 Measurement of surfactant and polymer adsorption

Surfactant and polymer adsorption are key to understanding how these molecules affect the stability/flocculation of the emulsion. Surfactant (both ionic and non-ionic) adsorption is reversible, and the process of adsorption can be described using the Gibbs isotherm, as described in Chapter 4. Basically, one measures the variation of interfacial tension with surfactant concentration C, and the amount of surfactant adsorption Γ can be calculated using the Gibbs equation [2]:

$$\Gamma_{2,1}^{\sigma} = -\frac{1}{RT}\left(\frac{d\gamma}{d\ln a_2^l}\right). \tag{13.7}$$

From Γ_∞ (the value at full saturation) the area per surfactant ion or molecule can be calculated:

$$\text{Area/molecule} = \frac{1}{\Gamma_\infty N_{av}}(\text{m}^2) = \frac{10^{18}}{\Gamma_\infty N_{av}}(\text{nm}^2). \tag{13.8}$$

As discussed in Chapter 4, the area per surfactant ion or molecule gives information on the orientation of surfactant ions or molecules at the interface. This information is relevant for the stability of the emulsion [3]. For example, for vertical orientation of surfactant ions, e.g. dodecyl sulphate anions, which is essential to produce a high surface charge (and hence enhanced electrostatic stability), the area per molecule is determined by the cross-sectional area of the sulphate group, which is in the region of 0.4 nm^2. With non-ionic surfactants consisting of an alkyl chain and poly(ethylene oxide) (PEO) head group adsorption on a hydrophobic surface is determined by the hydrophobic interaction between the alky chain and the hydrophobic surface. For vertical orientation of a mono-layer of surfactant molecules, the area per molecule depends on the size of the PEO chain. The latter is directly related to the number of EO units in the chain. If the area per molecule is smaller than that predicted from the size of the PEO chain, the surfactant molecules may associate on the surface forming bilayers, hemi-micelles. This information can be directly related to the stability of the emulsion.

The adsorption of polymers is more complex than surfactant adsorption, since one must consider the various interactions (chain-surface, chain-solvent and surface-solvent) as well as the conformation of the polymer chain on the surface [4]. Complete information on polymer adsorption may be obtained if one is able to determine the segment density distribution, i.e. the segment concentration in all layers parallel to the surface. However, such information is generally unavailable, and therefore one determines three main parameters: the amount of adsorption Γ per unit area, the fraction p of segments in direct contact with the surface (i.e. in trains) and the adsorbed layer thickness δ. Unfortunately such information is difficult to obtain at the liquid/liquid interface. However, low-angle neutron scattering can be applied to obtain this information. In this technique, deuterated oil is used, and the medium is made from a mixture of H_2O and D_2O. By adjusting the composition of the medium, one can match the scattering length density of the oil and dispersion medium. In this case, the scattering from the polymer layer can be obtained, which gives information on the thickness of the layer and its conformation [4].

13.3 Assessment of creaming/sedimentation of emulsions

As mentioned in Chapter 8, most emulsions undergo creaming or sedimentation on standing due to gravity and the density difference $\Delta\rho$ between the droplet and dispersion medium. This is particularly the case when the droplet radius exceeds 50 nm and when $\Delta\rho > 0.1$. In this case the Brownian diffusion cannot overcome the gravity force, and creaming or sedimentation occurs, resulting in an increase in droplet concentration from the bottom to the top of the container (for creaming) or from the top to the bottom (for sedimentation). As discussed in Chapter 8, to prevent droplet creaming or sedimentation, "thickeners" (rheology modifiers) are added in the continuous phase. The creaming/sedimentation of the emulsion is characterised by the creaming/sedimentation rate, cream/sediment volume, the change of droplet size distribution during creaming/settling and the stability of the emulsion to creaming/sedimentation. Assessment of creaming/sedimentation of an emulsion depends on the force applied to the droplets in the emulsion, namely gravitational, centrifugal and electrophoretic. The creaming/sedimentation processes are complex and subject to various errors in creaming/sedimentation measurements. An emulsion is usually agitated before measuring creaming/sedimentation, to ensure an initially homogeneous system of droplets in random motion. Vigorous agitation or the use of ultrasonic cavitation must be avoided to prevent any breakdown of aggregates and change of the droplet size distribution.

The practical measurement of creaming or sedimentation is hindered by the opacity of the emulsions. If there is any variation in the speed of the droplets due to polydispersity or density variation, the slower moving fraction obscures the movement of the faster droplets. Analysis of creaming or sedimentation rates needs a knowledge of the droplet concentration with height and time. Two methods can be applied to obtain such information, namely the use of back scattering of near infra-red (NIR). A schematic representation of an instrument that can be used for such measurement, namely the Turbiscan, is shown in Fig. 13.2. This technique consists in sending photons (light) into the sample. These photons, after being scattered by the emulsion droplets, emerge from the sample and are detected by the measurement device of the Turbiscan. A mobile reading head, composed of a NIR diode and two detectors (transition T) and back scattering BS, scans a cell containing the emulsion. The Turbiscan software then enables easy interpretation of the obtained data. The measurement en-

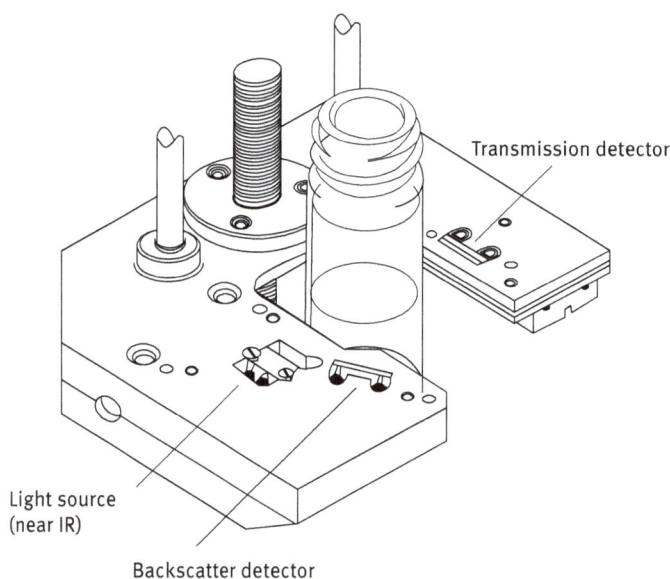

Fig. 13.2: Schematic representation of the Turbiscan.

ables the quantification of several parameters, as BS and T values are linked to droplet average diameter (d) and volume fraction (φ):

$$BS = f\left(\frac{d}{\varphi}\right). \tag{13.9}$$

A schematic representation of an ultrasonic method is shown in Fig. 13.3. The velocity of ultrasound through an emulsion is sensitive to composition [5]. This is the principle of the ultrasound monitor, shown in Fig. 13.3, which measures the ultrasonic velocity as a function of height. The time-of-flight of a pulse of ultrasound is measured across a rectangular sample cell immersed in thermostated water bath . The time of flight data are converted to ultrasonic velocity by reference to measurements made in two calibration fluids. The ultrasonic velocity data may be used to calculate the volume fraction of the oil using simple mixing theory.

The speed at which ultrasound propagates through an emulsion is a complex function of the droplet volume fraction, size and properties of the droplets and continuous phase. However, when the droplets are much smaller than the wavelength

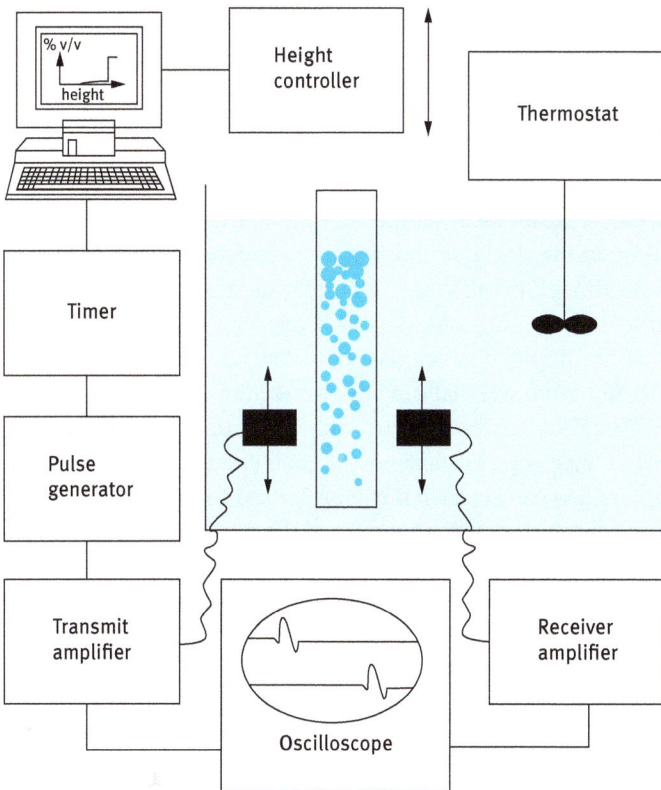

Fig. 13.3: Schematic representation of the ultrasonic creaming meter [5].

of ultrasound, and there is a significant difference between the speed of sound in the bulk dispersed and continuous phases, the effect of volume fraction greatly outweighs all other effects, so that the speed of ultrasound V may be calculated by assuming the system behaves like a simple mixture using the equation

$$V = \left(\frac{V_c^2}{\left(1 - \varphi\left(1 - \frac{\rho_d}{\rho_c}\right)\right)\left(1 - \varphi\left(1 - \frac{\rho_c V_c^2}{\rho_d V_d^2}\right)\right)} \right)^{1/2}, \tag{13.10}$$

where ρ_d, ρ_c, V_d and V_c are the densities and speeds of ultrasound through the dispersed and continuous phases respectively and φ is the volume fraction of the dispersed phase.

13.4 Assessment of flocculation, Ostwald ripening and coalescence

Assessment of flocculation, Ostwald ripening and coalescence of an emulsion requires measurement of the droplet size distribution as a function of time. Several techniques may be applied for this purpose, which are summarised below [6].

13.4.1 Optical microscopy

This is by far the most valuable tool for a qualitative or quantitative examination of the emulsion. Information on the size, morphology and aggregation of droplets can be conveniently obtained, with minimum time required for sample preparation. Since individual droplets can be directly observed, optical microscopy is considered to be the only absolute method for droplet characterisation. However, optical microscopy has some limitations: the minimum size that can be detected; the practical lower limit for accurate measurement of droplet size is 1.0 μm, although some detection may be obtained down to 0.3 μm. Image contrast may not be good enough for observation, particularly when using a video camera which is mostly used for convenience. The contrast can be improved by decreasing the aperture of the iris diaphragm, but this reduces the resolution. The contrast of the image depends on the refractive index of the particles relative to that of the medium. Hence the contrast can be improved by increasing the difference between the refractive index of the particles and the immersion medium. Unfortunately, changing the medium for the emulsion is not practical, since this may affect the state of the emulsion. Fortunately, water with a refractive index of 1.33 is a suitable medium for most organic droplets, with a refractive index usually > 1.4.

The ultramicroscope by virtue of dark field illumination extends the useful range of optical microscopy to small droplets not visible in a bright light illumination. Dark field illumination utilizes a hollow cone of light at a large angle of incidence. The image is formed by light scattered from the droplets against a dark background. Droplets about 10 times smaller than those visible under bright light illumination can be detected. However, the image obtained is abnormal, and the droplet size cannot be accurately measured. Three main attachments to the optical microscope are possible:

Phase contrast
This utilizes the difference between the diffracted waves from the main image and the direct light from the light source. The specimen is illuminated with a light cone, and this illumination is within the objective aperture. The light illuminates the specimen and generates zero order and higher orders of diffracted light. The zero-order light beam passes through the objective and a phase plate which is located at the objective back focal plane. The difference between the optical path of the direct light beam and that of the beam diffracted by a droplet causes a phase difference. The constructive and destructive interferences result in brightness changes which enhance the contrast. This produces sharp images, allowing one to obtain droplet size measurements more accurately. The phase contrast microscope has a plate in the focal plane of the objective back focus. The condenser is equipped instead of a conventional iris diaphragm with a ring matched in its dimension to the phase plate.

Differential interference contrast (DIC)
This gives a better contrast than the phase contrast method. It utilizes a phase difference to improve contrast, but the separation and recombination of a light beam into two beams is accomplished by prisms. DIC generates interference colours, and the contrast effects indicate the refractive index difference between the droplet and medium.

Polarized light microscopy
This illuminates the sample with linearly or circularly polarized light, either in a reflection or transmission mode. One polarizing element, located below the stage of the microscope, converts the illumination into polarized light. The second polarizer is located between the objective and the ocular and is used to detect polarized light. Linearly polarized light cannot pass the second polarizer in a crossed position unless the plane of polarization has been rotated by the specimen.

Sample preparation for optical microscopy
A drop of the emulsion is placed on a glass slide and covered with a cover glass. If the emulsion has to be diluted, the dispersion medium (which can be obtained by

centrifugation of the emulsion) should be used as the diluent in order to avoid aggregation. At low magnifications the distance between the objective and the sample is usually adequate for manipulating the sample, but at high magnification the objective may be too close to the sample. An adequate working distance can be obtained while maintaining high magnification by using a more powerful eyepiece with a low power objective. For emulsions encountering Brownian motion (when the droplet size is relatively small), microscopic examination of moving droplets can become difficult. In this case one can record the image on a photographic film or video tape or disc (using computer software).

Droplet size measurements using optical microscopy

The optical microscope can be used to observe dispersed droplets and flocs. Droplet sizing can be carried out using manual, semi-automatic or automatic image analysis techniques. In the manual method (which is tedious) the microscope is fitted with a minimum of $10 \times$ and $43 \times$ achromatic or apochromatic objectives equipped with a high numerical apertures ($10 \times$, $15 \times$ and $20 \times$), a mechanical XY stage, a stage micrometer and a light source. The direct measurement of droplet size is aided by a linear scale or globe-and-circle graticules in the ocular. The linear scale is useful mainly for spherical droplets, with a relatively narrow droplet size distribution. The globe-and-circle graticules are used to compare the projected droplet area with a series of circles in the ocular graticule. The size of spherical droplets can be expressed by the diameter. One of the difficulties with the evaluation of emulsions by optical microscopy is the quantification of data. The number of droplets in at least six different size ranges must be counted to obtain a distribution. This problem can be alleviated by the use of automatic image analysis which can also give an indication on the floc size and its morphology.

13.4.2 Electron microscopy

Electron microscopy utilizes an electron beam to illuminate the sample. The electrons behave as charged particles which can be focused by annular electrostatic or electromagnetic fields surrounding the electron beam. Due to the very short wavelength of electrons, the resolving power of an electron microscope exceeds that of an optical microscope by ~ 200 times. The resolution depends on the accelerating voltage which determines the wavelength of the electron beam, and magnifications as high as 200 000 can be reached with intense beams, but this could damage the sample. Mostly the accelerating voltage is kept below 100–200 kV, and the maximum magnification obtained is below 100 000. The main advantage of electron microscopy is the high resolution, sufficient for resolving details separated by only a fraction of a nanometer. The increased depth of field, usually by about 10 μm or about 10 times that of an optical microscope, is another important advantage of electron microscopy. Nevertheless, electron microscopy has also some disadvantages, such as sample preparation, selection of the area viewed and interpretation of the data. The main drawback of electron microscopy is the potential risk of altering or damaging the sample that may introduce artifacts and possible aggregation of the droplets during sample preparation. The emulsion has to be frozen, and the removal of the dispersion medium may alter the distribution of the droplets. If the droplets do not conduct electricity, the sample has to be coated with a conducting layer, such as gold, carbon or platinum, to avoid negative charging by the electron beam. Two main types of electron microscopes are used: transmission and scanning.

Transmission electron microscopy (TEM)

TEM displays an image of the specimen on a fluorescent screen, and the image can be recorded on a photographic plate or film. The TEM can be used to examine droplets in the range 0.001–5 µm. The sample is deposited on a Formvar (polyvinyl formal) film resting on a grid to prevent charging of the simple. The sample is usually observed as a replica by coating with an electron transparent material (such as gold or graphite). The preparation of the sample for the TEM may alter the state of emulsion and cause aggregation. Freeze fracturing techniques have been developed to avoid some of the alterations of the sample during sample preparation. Freeze fracturing allows the emulsion to be examined without dilution, and replicas can be made of emulsions containing water. It is necessary to have a high cooling rate to avoid the formation of ice crystals.

Scanning electron microscopy (SEM)

SEM can show droplet topography by scanning a very narrowly focused beam across the droplet surface. The electron beam is directed normally or obliquely at the surface. The back-scattered or secondary electrons are detected in a raster pattern and displayed on a monitor screen. The image provided by secondary electrons exhibit good three-dimensional detail. The back-scattered electrons, reflected from the incoming electron beam, indicate regions of high electron density. Most SEMs are equipped with both types of detectors. The resolution of the SEM depends on the energy of the electron beam which does not exceed 30 kV, and hence the resolution is lower than that obtained by the TEM. A very important advantage of SEM is elemental analysis by energy dispersive x-ray analysis (EDX). If the electron beam impinging on the specimen has sufficient energy to excite atoms on the surface, the sample will emit x-rays. The energy required for x-ray emission is characteristic of a given element, and since the emission is related to the number of atoms present, quantitative determination is possible.

13.4.3 Confocal laser scanning microscopy (CLSM)

CLSM is a very useful technique for identification of emulsions. It uses a variable pin-hole aperture or variable width slit to illuminate only the focal plane by the apex of a cone of laser light. Out-of-focus items are dark and do not distract from the contrast of the image. As a result of extreme depth discrimination (optical sectioning) the resolution is considerably improved (up to 40 % when compared with optical microscopy). The CLSM technique acquires images by laser scanning or uses computer software to subtract out-of-focus details from the in-focus image. Images are stored as the sample is advanced through the focal plane in elements as small as 50 nm. Three-dimensional images can be constructed to show the shape of the droplets.

13.5 Scattering techniques

These are by far the most useful methods for characterisation of emulsions, and in principle they can give quantitative information on the droplet size distribution, floc size and shape. The only limitation of the methods is the need to use sufficiently dilute samples to avoid interference such as multiple scattering, which makes interpretation of the results difficult. However, recently back-scattering methods have been designed which allow measurement of a sample without dilution. In principle one can use any electromagnetic radiation such as light, x-ray or neutrons, but in most industrial labs only light scattering is applied (using lasers).

13.5.1 Light-scattering techniques

These can be conveniently divided into the following classes [6]: (i) time-average light scattering; Static or elastic scattering; (ii) turbidity measurements which can be carried out using a simple spectrophotometer; (iii) light diffraction technique; (iv) dynamic (quasi-elastic) light scattering that is usually referred as photon correlation spectroscopy. This is a rapid technique that is very suitable for measuring submicron particles or droplets (nano-size range); (v) back scattering techniques that is suitable for measuring concentrated samples. Application of any of these methods depends on the information required and availability of the instrument.

Time-average light scattering
In this method the dispersion that is sufficiently diluted to avoid multiple scattering is illuminated by a collimated light (usually laser) beam, and the time-average intensity of scattered light is measured as a function of scattering angle θ. Static light scattering is termed elastic scattering. Three regimes can be identified:

Rayleigh regime

Whereby the particle radius R is smaller than $\lambda/20$ (where λ is the wave length of incident light). The scattering intensity is given by the equation

$$I(Q) = [\text{Instrument constant}][\text{Material constant}]\, NV_p^2, \qquad (13.11)$$

where Q is the scattering vector that depends on the wavelength of light λ used and is given by

$$Q = \left(\frac{4\pi n}{\lambda}\right) \sin\left(\frac{\theta}{2}\right), \qquad (13.12)$$

where n is the refractive index of the medium.

The material constant depends on the difference between the refractive index of the droplet and that of the medium. N is the number of droplets and V_p is the volume of each droplet. Assuming that the droplets are spherical one can obtain the average size using equation (13.11).

The Rayleigh equation reveals two important relationships: (i) The intensity of scattered light increases with the square of the droplet volume and consequently with the sixth power of the radius R. Hence the scattering from larger droplets may dominate the scattering from smaller droplets. (ii) The intensity of scattering is inversely proportional to λ^4. Hence a decrease in the wavelength will substantially increase the scattering intensity.

Rayleigh–Gans–Debye regime (RGD) $\lambda/20 < R < \lambda$

The RGD regime is more complicated than the Rayleigh regime, and the scattering pattern is no longer symmetrical about the line corresponding to the 90° angle, but favors forward scattering ($\theta < 90°$) or back scattering ($180° > \theta > 90°$). Since the preference of forward scattering increases with increasing droplet size, the ratio $I_{45°}/I_{135°}$ can indicate the droplet size.

Mie regime $R > \lambda$

The scattering behaviour is more complex than the RGD regime, and the intensity exhibits maxima and minima at various scattering angles depending on droplet size and refractive index. The Mie theory for light scattering can be used to obtain the droplet size distribution using numerical solutions. One can also obtain information on droplet shape.

13.5.2 Turbidity measurements

Turbidity (total light-scattering technique) can be used to measure droplet size, flocculation and droplet creaming/sedimentation. This technique is simple and easy to use; a single or double beam spectrophotometer or a nephelometer can be used.

For non-absorbing droplets the turbidity τ is given by

$$\tau = (1/L) \ln (I_0/I) , \tag{13.13}$$

where L is the path length, I_0 is the intensity of incident beam and I is the intensity of transmitted beam.

The droplet size measurement assumes that the light scattering by a droplet is singular and independent on other particles Any multiple scattering complicates the analysis. According to the Mie theory, the turbidity is related to the droplet number N and their cross section πr^2 (where r is the droplet radius) by

$$\tau = Q \pi r^2 N , \tag{13.14}$$

where Q is the total Mie scattering coefficient. Q depends on the droplet size parameter α (which depends on droplet diameter and wave length of incident light λ) and the ratio of refractive index of the droplets and medium m. Q depends on α in an oscillatory mode, and it exhibits a series of maxima and minima whose position depends on m. For droplets with $R < \frac{1}{20}\lambda$, $\alpha < 1$, and it can be calculated using the Rayleigh theory. For $R > \lambda$, Q approaches 2, and between these two extremes the Mie theory is used. If the droplets are not mono-disperse (as is the case with most practical systems), the droplet size distribution must be taken into account. Using this analysis one can establish the droplet size distribution using numerical solutions.

13.5.3 Light diffraction techniques

This is a rapid and non-intrusive technique for determination of droplet size distribution in the range 2–300 μm, with good accuracy for most practical purposes. Light diffraction gives an average diameter over all droplet orientations, as randomly oriented droplets pass the light beam. A collimated and vertically polarized laser beam illuminates droplet dispersion and generates a diffraction pattern with the undiffracted beam in the centre. The energy distribution of diffracted light is measured by a detector consisting of light sensitive circles separated by isolating circles of equal width. The angle formed by the diffracted light increases with decreasing particle size. The angle-dependent intensity distribution is converted by Fourier optics into a spatial intensity distribution I(r). The spatial intensity distribution is converted into a set of photocurrents, and the droplet size distribution is calculated using a computer. Several commercial instruments are available, e.g. Malvern Master Sizer (Malvern, UK), Horriba (Japan) and Coulter LS Sizer (USA). A schematic illustration of the set-up is shown in Fig. 13.4.

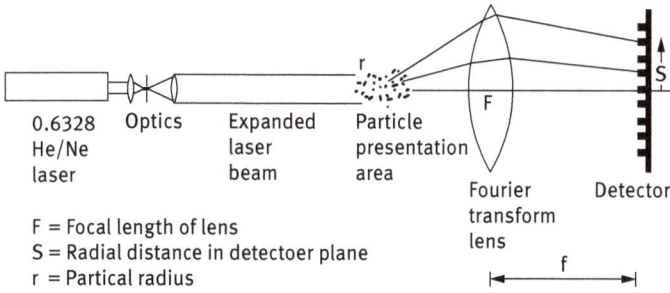

F = Focal length of lens
S = Radial distance in detectoer plane
r = Partical radius

Fig. 13.4: Schematic illustration of light diffraction particle sizing system.

In accordance with the Fraunhofer theory (which was introduced by Fraunhofer over 100 years ago), the special intensity distribution is given by

$$I(r) = \int_{X_{min}}^{X_{max}} N_{tot} q_0(x) I(r, x) \, dx, \tag{13.15}$$

where $I(r, x)$ is the radial intensity distribution at radius r for droplets of size x, N_{tot} is the total number of droplets and $q_0(x)$ describes the droplet size distribution.

The radial intensity distribution $I(r, x)$ is given by

$$I(r, x) = I_0 \left(\frac{\pi x^2}{2f} \right)^2 \left(\frac{J_i(k)}{k} \right)^2, \tag{13.16}$$

with $k = (\pi x r)/(\lambda f)$, and where r is the distance to the centre of the disc, λ is the wavelength, f is the focal length and J_i is the first-order Bessel function.

The Fraunhofer diffraction theory applies to droplets whose diameter is considerably larger than the wavelength of illumination. As shown in Fig. 13.2, a He/Ne laser is used with $\lambda = 632.8$ nm for droplet sizes mainly in the 2–120 μm range. In general the diameter of the sphere-shaped droplet should be at least four times the wavelength of the illumination light. The accuracy of droplet size distribution determined by light diffraction is not very good if a large fraction of droplets with diameter <10 μm is present in the emulsion. For small droplets (diameter < 10 μm) the Mie theory is more accurate if the necessary optical parameters, such as refractive index of droplets and medium and the light absorptivity of the dispersed droplets, are known. Most commercial instruments combine light diffraction with forward light scattering to obtain a full droplet size distribution covering a wide range of sizes.

Fig. 13.5 shows the result of particle sizing using a six component mixture of standard polystyrene lattices (using a Mastersizer).

Most practical emulsions are polydisperse and generate a very complex diffraction pattern. The diffraction pattern of each droplet size overlaps with diffraction patterns of other sizes. The droplets of different sizes diffract light at different angles, and the

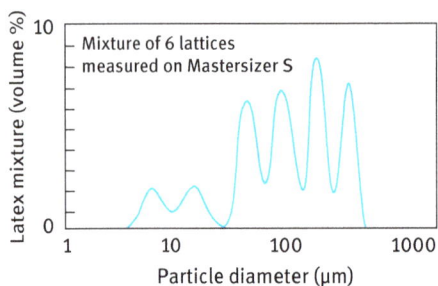

Fig. 13.5: Single measurement of a mixture of six standard lattices using the Mastersizer.

energy distribution becomes a very complex pattern. However, manufacturers of light diffraction instruments (such as Malvern, Coulters and Horriba) developed numerical algorithms relating diffraction patterns to droplet size distribution.

Several factors can affect the accuracy of Fraunhofer diffraction: (i) droplets smaller than the lower limit of Fraunhofer theory; (ii) non-existent "ghost" droplets in droplet size distribution obtained by Fraunhofer diffraction applied to systems containing droplets with a large fraction of small sizes (below 10 μm); (iii) computer algorithms that are unknown to the user and vary with the manufacturer software version; (iv) the composition-dependent optical properties of the droplets and dispersion medium. If the density of all droplets is not the same, the result may be inaccurate.

13.5.4 Dynamic light scattering – photon correlation spectroscopy (PCS)

Dynamic light scattering (DLS) is a method that measures the time-dependent fluctuation of scattered intensity. It is also referred to as quasi-elastic light scattering (QELS), or photon correlation spectroscopy (PCS). The latter is the most commonly used term for describing the process, since most dynamic scattering techniques employ autocorrelation. PCS is a technique that utilizes the Brownian motion to measure the droplet size. As a result of Brownian motion of dispersed droplets the intensity of scattered light undergoes fluctuations which are related to the velocity of the droplets. Since larger droplets move less rapidly than the smaller ones, the intensity fluctuation (intensity vs time) pattern depends on droplet size, as illustrated in Fig. 13.6. The velocity of the scatterer is measured in order to obtain the diffusion coefficient.

In a system where the Brownian motion is not interrupted by creaming/sedimentation or droplet-droplet interaction, the movement of droplets is random. Hence, the intensity fluctuations observed after a large time interval do not resemble those fluctuations observed initially, but represent a random distribution of droplets. Consequently the fluctuations observed at large time delay are not correlated with the initial fluctuation pattern. However, when the time differential between the observations is very small (a nano- or micro-second), both positions of droplets are similar, and the scattered intensities are correlated. When the time interval is increased, the correla-

Large particles

Small particles

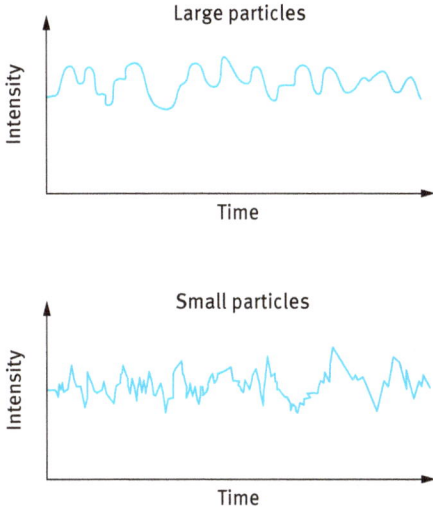

Fig. 13.6: Schematic representation of the intensity fluctuation for large and small particles.

tion decreases. The decay of correlation is particle size dependent. The smaller the particles are, the faster is the decay.

The fluctuations in scattered light are detected by a photomultiplier and are recorded. The data containing information on the droplet motion are processed by a digital correlator. The latter compares the intensity of scattered light at time t, I(t), to the intensity at a very small time interval τ later, I(t + τ). and it constructs the second-order auto-correlation function $G_2(\tau)$ of the scattered intensity:

$$G_2(\tau) = \langle I(t)I(t + \tau)\rangle \,. \tag{13.17}$$

The experimentally measured intensity autocorrelation function $G_2(\tau)$ depends only on the time interval τ, and is independent of t, the time when the measurement started.

PCS can be measured in a homodyne, where only scattered light is directed to the detector. It can also be measured in heterodyne mode, where a reference beam split from the incident beam is superimposed on scattered light. The diverted light beam functions as a reference for the scattered light from each particle.

In homodyne mode, $G_2(\tau)$ can be related to the normalized field autocorrelation function $g_1(\tau)$ by

$$G_2(\tau) = A + Bg_1^2(\tau)\,, \tag{13.18}$$

where A is the background term designated as the baseline value and B is an instrument-dependent factor. The ratio B/A is regarded as a quality factor of the measurement or the signal-to-noise ratio and expressed sometimes as the % merit.

The field autocorrelation function $g_1(\tau)$ for a monodisperse emulsion decays exponentially with τ:

$$g_1(\tau) = \exp(-\Gamma\tau), \qquad (13.19)$$

where Γ is the decay constant (s^{-1}).

Substituting equation (13.19) into equation (13.18) yields the measured autocorrelation function

$$G_2(\tau) = A + B\exp(-2\Gamma\tau). \qquad (13.20)$$

The decay constant Γ is linearly related to the translational diffusion coefficient D_T of the droplet:

$$\Gamma = D_T q^2. \qquad (13.21)$$

The modulus q of the scattering vector is given by

$$q = \frac{4\pi n}{\lambda_o}\sin\left(\frac{\theta}{2}\right), \qquad (13.22)$$

where n is the refractive index of the dispersion medium, θ is the scattering angle and λ_o is the wavelength of the incident light in vacuum.

PCS determines the diffusion coefficient, and the droplet radius R is obtained using the Stokes–Einstein equation:

$$D = \frac{kT}{6\pi\eta R}, \qquad (13.23)$$

where k is the Boltzmann constant, T is the absolute temperature and η is the viscosity of the medium.

The Stokes–Einstein equation is limited to non-interacting, spherical and rigid spheres. The effect of droplet interaction at relatively low droplet concentration c can be taken into account by expanding the diffusion coefficient into a power series of concentration:

$$D = D_0(1 + k_D c), \qquad (13.24)$$

where D_o is the diffusion coefficient at infinite dilution and k_D is the virial coefficient that is related to droplet interaction. D_o can be obtained by measuring D at several droplet number concentrations and extrapolating to zero concentration.

For polydisperse emulsion the first-order auto-correlation function is an intensity-weighted sum of the auto-correlation function of particles contributing to the scattering:

$$g_1(\tau) = \int_0^\infty C(\Gamma)\exp(-\Gamma\tau)\,d\Gamma. \qquad (13.25)$$

$C(\Gamma)$ represents the distribution of decay rates.

For narrow droplet size distribution the cumulant analysis is usually satisfactory The cumulant method is based on the assumption that for mono-disperse emulsions

$g_1(\tau)$ is mono-exponential. Hence $\log g_1(\tau)$ vs τ yields a straight line with a slope equal to Γ:

$$\ln g_1(\tau) = 0.5 \ln (B) - \Gamma\tau, \tag{13.26}$$

where B is the signal-to-noise ratio.

The cumulant method expands the Laplace transform about an average decay rate:

$$\langle\Gamma\rangle = \int_0^\infty \Gamma C(\Gamma)\,d\Gamma. \tag{13.27}$$

The exponential in equation (13.26) is expanded by an average and integrated term:

$$\ln g_1(\tau) = \langle\Gamma\rangle\tau + (\mu_2\tau^2)/2! - (\mu_3\tau^3)/3! + \ldots \tag{13.28}$$

An average diffusion coefficient is calculated from $\langle\Gamma\rangle$, and the polydispersity (termed the polydispersity index) is indicated by the relative second moment, $\mu_2/\langle\Gamma\rangle^2$. A constrained regulation method (CONTIN) yields several numerical solutions to the droplet size distribution, and this is normally included in the software of the PCS machine.

PCS is a rapid, absolute, non-destructive and rapid method for droplet size measurements. It has some limitations. The main disadvantage is the poor resolution of the droplet size distribution. Also it suffers from the limited size range (absence of any creaming/sedimentation) that can be accurately measured. Several instruments are commercially available, e.g. by Malvern, Brookhaven, Coulters, etc.

13.5.5 Back-scattering techniques

This method is based on the use of fibreoptics, sometimes referred to as fibreoptic dynamic light scattering (FODLS), and it allows one to measure at high droplet number concentrations. The FODLS employ either one or two optical fibres. Alternatively, fibre bundles may be used. The exit port of the optical fibre (optode) is immersed in the sample, and the scattered light in the same fibre is detected at a scattering angle of 180° (i.e. back scattering).

The above technique is suitable for online measurements during manufacture of an emulsion. Several commercial instruments are available, e.g. Lesentech (USA).

13.6 Measurement of the rate of creaming or sedimentation

This was discussed previously, using either the Turbiscan or ultrasound measurements.

13.7 Measurement of rate of flocculation

Two general techniques may be applied for measuring the rate of flocculation of emulsions, both of which can only be applied for dilute systems. The first method is based on measuring the scattering of light by the particles. For monodisperse droplets with a radius that is less than $\lambda/20$ (where λ is the wave length of light) one can apply the Rayleigh equation, whereby the turbidity τ_0 is given by

$$\tau_0 = A'n_0V_1^2,\tag{13.29}$$

where A' is an optical constant (which is related to the refractive index of the particle and medium and the wave length of light) and n_0 is the number of droplets, each with a volume V_1. By combining the Rayleigh theory with the Smoluchowski–Fuchs theory of flocculation kinetics [7, 8], one can obtain the following expression for the variation of turbidity with time:

$$\tau = A'n_0V_1^2(1 + 2n_0kt),\tag{13.30}$$

where k is the rate constant of flocculation

The second method for obtaining the rate constant of flocculation is by direct droplet counting as a function of time. For this purpose optical microscopy or image analysis may be used, provided the droplet size is within the resolution limit of the microscope. Alternatively, the droplet number may be determined using electronic devices such as the Coulter counter or the flow ultramicroscope.

The rate constant of flocculation is determined by plotting $1/n$ versus t, where n is the number of particles after time t, i.e.

$$\left(\frac{1}{n}\right) = \left(\frac{1}{n_0}\right) + kt.\tag{13.31}$$

The rate constant k of slow flocculation is usually related to the rapid rate constant k_0 (the Smoluchowski rate) by the stability ratio W:

$$W = \left(\frac{k}{k_0}\right).\tag{13.32}$$

One usually plots $\log W$ versus $\log C$ (where C is the electrolyte concentration) to obtain the critical coagulation concentration (c.c.c.), which is the point at which $\log W = 0$.

A very useful method for measuring flocculation is to use the single-droplet optical method, using the droplets of the emulsion that are dispersed in a liquid flow through a narrow uniformly illuminated cell. The emulsion is made sufficiently dilute (using the continuous medium) so that droplets pass through the cell individually. A droplet passing through the light beam illuminating the cell generates an optical pulse detected by a sensor. If the droplet size is greater than the wavelength of light ($> 0.5\,\mu m$), the peak height depends on the projected area of the droplet. If the droplet size is smaller than 0.5 μm, the scattering dominates the response. For droplets > 1 μm, a light obscuration (also called blockage or extinction) sensor is used. For droplets smaller than 1 μm, a light-scattering sensor is more sensitive.

The above method can be used to determine the size distribution of aggregating emulsions. The aggregated droplets pass individually through the illuminated zone and generate a pulse which is collected at a small angle (< 3°). At sufficiently small angles, the pulse height is proportional to the square of the number of monomeric units in an aggregate and independent of the aggregate shape or its orientation.

13.8 Measurement of incipient flocculation

This can be done for sterically stabilised suspensions, when the medium for the chains becomes a θ-solvent. This occurs, for example, on heating an aqueous emulsion stabilised with poly(ethylene oxide) (PEO) or poly(vinyl alcohol) chains. Above a certain temperature (the θ-temperature), which depends on electrolyte concentration, flocculation of the emulsion occurs. The temperature at which this occurs is defined as the critical flocculation temperature (CFT).

This process of incipient flocculation can be followed by measuring the turbidity of the emulsion as a function of temperature. Above the CFT, the turbidity of the emulsion rises very sharply.

For the above purpose, the cell in the spectrophotometer used to measure the turbidity is placed in a metal block connected to a temperature programming unit (which allows one to increase the temperature rise at a controlled rate).

13.9 Measurement of Ostwald ripening

As discussed in Chapter 10, Ostwald ripening is the result of the difference in solubility S between small and large droplets. The smaller droplets have larger solubility than the larger particles. The effect of droplet size on solubility is described by the Kelvin equation [9]:

$$S(r) = S(\infty) \exp \left(\frac{2\gamma V_m}{rRT} \right) \tag{13.33}$$

where $S(r)$ is the solubility of a droplet with radius r, $S(\infty)$ is the solubility of a droplet with infinite radius, γ is the liquid/liquid interfacial tension, V_m is the molar volume of the disperse phase (= M/ρ, where M is the molecular weight and ρ is the density of the droplets), R is the gas constant and T is the absolute temperature.

For two droplets with radii r_1 and r_2,

$$\frac{RT}{V_m} \ln \left(\frac{S_1}{S_2} \right) = 2\gamma \left(\frac{1}{r_1} - \frac{1}{r_2} \right). \tag{13.34}$$

To obtain a measure of the rate of crystal growth, the droplet size distribution of the emulsion is followed as a function of time, using either a Coulter counter, a Mastersizer or an optical disc centrifuge. One usually plots the cube of the average radius vs time, which gives a straight line from which the rate of Ostwald ripening can be determined (the slope of the linear curve):

$$r^3 = \frac{8}{9}\left[\frac{S(\infty)\gamma V_m D}{\rho RT}\right]t.$$

(13.35)

D is the diffusion coefficient of the disperse phase in the continuous phase and ρ is the density of the droplets.

13.10 Measurement of the rate of coalescence

As discussed in Chapter 11, coalescence is the combination of two or more droplets to form a larger drop. The rate of coalescence can be expressed in terms of a first-order rate equation [10, 11] if the droplets will have flocculated in a time which is much shorter than the time scale of coalescence. If K (s^{-1}) is the coalescence rate constant, then

$$-\frac{dn}{dt} = Kn,$$

(13.36)

where n is the number of droplets at time t; or

$$n = n_0 \exp(-Kt),$$

(13.37)

where n_0 is the number of droplets at t = 0.

Equation (13.37) shows that a plot of log n vs t gives a straight line, and the slope is equal to K. The number of droplets at any time in an emulsion can be measured using a Coulter counter. Alternatively on can measure the average diameter d of the droplets as a function of time using the Mastersizer described above:

$$d = d_0 \exp(Kt).$$

(13.38)

Again a plot of log d versus time gives a straight line with a positive slope equal to K.

13.11 Bulk properties of emulsions. Equilibrium cream or sediment volume (or height)

For a "structured" emulsion, obtained by "controlled flocculation" or addition of "thickeners" (such polysaccarides, clays or oxides), the "flocs" cream or sediment at a rate depending on their size and porosity of the aggregated structure. After this initial creaming or sedimentation, compaction and rearrangement of the floc structure occurs, a phenomenon referred to as consolidation.

Normally in cream or sediment volume measurements, one compares the initial volume V_0 (or height H_0) with the ultimately reached value V (or H). A colloidally stable emulsion gives a "close-packed" structure with relatively small cream or sediment volume. A weakly "flocculated" or "structured" emulsion gives a more open cream or sediment and hence a higher cream or sediment volume. Thus, by comparing the relative sediment volume V/V_0 or height H/H_0, one can distinguish between a colloidally stable and flocculated emulsion.

References

[1]   R. J. Hunter, *Zeta Potential in Colloid Science: Principles and Application*, Academic Press, London, 1981.

[2]   J. W. Gibbs, *Collected Works,* vol. 1, p. 219, Longman, New York, 1928.

[3]   M. J. Rosen and J. T. Kunjappu, *Surfactants and Interfacial Phenomena*, John Wiley and Sons, New Jersey, 2012.

[4]   G. J. Fleer, M. A. Cohen-Stuart, J. M. H. M. Scheutjens, T. Cosgrove and B. Vincent, *Polymers at Interfaces*, Chapman and Hall, London, 1993.

[5]   M. M. Robins and D. J. Hibberd, Emulsion Flocculation and Creaming, in: B. P. Binks (ed.), *Modern Aspects of Emulsion Science*, Ch. 4, Royal Society of Chemistry, Cambridge, 1998.

[6]   E. Kissa, *Dispersions, Characterization, Testing and Measurement*, Marcel Dekker, New York, 1999.

[7]   M. Von Smoluchowski, *Physik. Z.* **17** (1916), 557, 585.

[8]   N. Fuchs, *Z. Physik* **89** (1936), 736.

[9]   W. Thompson (Lord Kelvin), *Phil. Mag.* **42** (1871) , 448.

[10] Th. F. Tadros and B. Vincent, in: P. Becher (ed.), *Encyclopedia of Emulsion Technology*, Marcel Dekker, New York, 1983.

[11] M. van den Tempel, *Rec. Trav. Chim.*, **72** (1953), 433, 442.

14 Industrial applications of emulsions

14.1 Introduction

Emulsions are widely used in many industrial applications, the following of which are worth mentioning: food (which emulsions are by far the most widely used in systems such as mayonnaise, salad creams, beverages, etc.); cosmetic and personal care products, such as hand creams, lotions, sunscreens, hair sprays, etc.); pharmaceuticals; agrochemicals (for formulation of many herbicides, insecticides, plant growth regulators, etc.); rolling oils and lubricants, etc. In the following we give a description of some of these applications.

14.2 Food emulsions

Many food emulsions are colloidal systems, containing droplets of various kinds that are stabilised by surfactants [1]. The interfacial properties of these surfactant films is very important in formulating such systems and maintaining their long-term physical stability. Naturally occurring surfactants such as lecithin from egg yolk and various proteins from milk are used for the preparation of many food emulsions such as mayonnaise, salad creams, dressings, deserts etc. Later, polar lipids such as monoglycerides were introduced as emulsifiers for food emulsions. More recently, synthetic surfactants such as sorbitan esters and their ethoxylates and sucrose esters have been used in food emulsions. For example, esters of monostearate or mono-oleate with organic carboxylic acids, e.g. citric acid, are used as anti-spattering agents in margarine for frying. The droplets may remain as individual units suspended in the medium, but in most cases aggregation of these droplets takes place forming three-dimensional structures, generally referred to as "gels". These aggregation structures are determined by the interfacial properties of the surfactant films and the interaction forces between the droplets that are controlled by the relative magnitudes of attractive (van der Waals forces) and repulsive forces. The latter can be electrostatic or steric in nature, depending on the composition of the food formulation. It is clear that the repulsive interactions will be determined by the nature of the surfactant present in the formulation. Such surfactants can be ionic or polar in nature, or they may be polymeric in nature. The latter are sometimes added not only to control the interaction between droplets in the food formulation, but also to control the consistency (rheology) of the system. Many food formulations contain mixtures of surfactants (emulsifiers) and hydrocolloids. The interaction between the surfactant and polymer molecule plays a major role in the overall interaction between the droplets, as well as the bulk rheology of the whole system. Such interactions are complex and require fundamental studies of their colloidal properties. As will be discussed later, many food emulsions contain proteins [2] used as emulsifiers. The interaction between proteins and hydrocolloids is also very important in determining the interfacial properties and bulk rheology of the system. In addition, the proteins can also interact with the emulsifiers present in the system, and this interaction requires particular attention. Surfactant association structures and emulsions play an important role in the food industry [3]

14.2.1 Food grade surfactants

Food grade surfactants are, in general, not soluble in water, but they can form association structures in aqueous media that are liquid crystalline in nature [1]. Three main liquid crystalline structures may be distinguished, namely the lamellar phase, the hexagonal phase and the cubic phase. Fig. 14.1 shows a model of the crystalline state of a surfactant which forms a lamellar phase (Fig. 14.1 (a)).

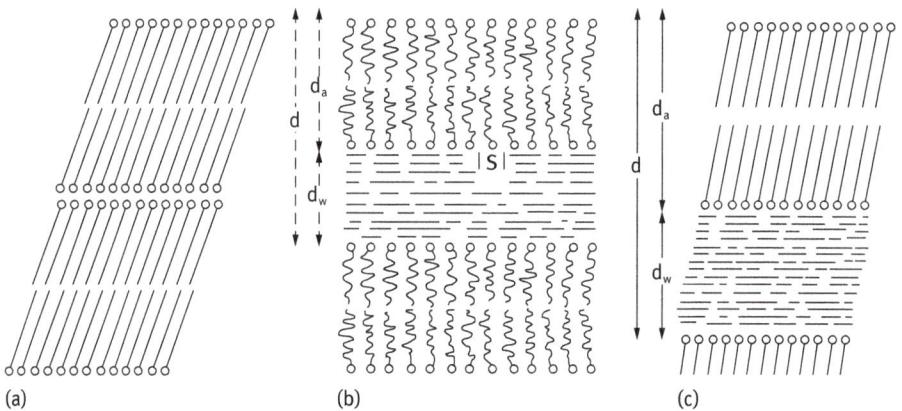

(a)   (b)   (c)

Fig. 14.1: Schematic representation of lamellar liquid crystalline structures.

When dispersed in water above its Krafft temperature (T_c) it produces a lamellar mesophase (Fig. 14.1 (b)) with a thickness d_a of the bilayer, a thickness d_w of the water layer. The lamellar layer thickness d is simply $d_a + d_w$. These thicknesses can be determined using low angle x-ray diffraction. The surface area per molecule of surfactant is denoted by S. The lamellar mesophase can be diluted with water and has almost infinite swelling capacity provided the lipid bilayers contain charged molecules and the water phase has a low ion concentration [4]. These diluted lamellar phases may form liposomes (multi-lamellar vesicles), which are spherical aggregates with internal lamellar structures [5]. Under the polarising microscope, the lamellar structures show "oil-streaky" texture.

When the surfactant solution containing the lamellar phase is cooled below the Krafft temperature of the surfactant, a gel phase is formed, as schematically shown in Fig. 14.1 (c). The crystalline structure of the bilayer is now similar to that of the pure surfactant, and the aqueous layer with thickness dw is the continuous phase of the gel.

The hexagonal mesophase structure is periodic in two dimensions and exists in two modifications: hexagonal I and hexagonal II. The structure of the hexagonal I phase consists of cylindrical aggregates of surfactant molecules with the polar head groups oriented towards the outer (continuous) water phase and the surfactant hydrocarbon chains filling out the core of the cylinders. These structures show a fan-like or angular texture under the polarising microscope (see section on concentrated surfactant solutions). The hexagonal II phase consists of cylindrical aggregates of water in a continuous medium of surfactant molecules with the polar head groups oriented towards the water phase and the hydrocarbon chains filling out the exterior between the water cylinders Under the polarising microscope, this phase shows the same angular texture as the hexagonal I phase. Whereas the hexagonal I phase can be diluted with water to produce micellar (spherical) solutions, the hexagonal II phase has a limited swelling capacity (usually not more than 40 % water in the cylindrical aggregates).

The viscous isotropic cubic phase, which is periodic in three dimensions, is produced with monogylceride-water systems at chain lengths above C_{14}. This isotropic phase was shown to consist of a bicontinuous structure, consisting of a lamellar bilayer, which separates two water channel systems [6, 7]. The cubic phase behaves as a very viscous liquid phase, which can accommodate up to ~ 40 % water.

Of the above liquid crystalline structures, the lamellar phase is the most important for food applications. As we will see later, these lamellar structures are very good stabilisers for food emulsions. In addition, they can be diluted with water forming liposome dispersions, which are easy to handle (pumpable liquids), and they interact with water soluble components such as amylose in starch particles. The hexagonal and cubic phases, in contrast, when formed present problems in food processing, due to their highly viscous nature (viscous particles may block filters).

The phase diagram of a pure soybean lecithin-water system is shown in Fig. 14.2. The excess water region, relevant to emulsions based on this surfactant, consists of a dispersion of the lamellar liquid crystalline phase in the form of liposomes.

A typical ternary phase diagram of soybean oil (triglyceride), sunflower oil monoglyceride and water at 25 °C [8] is shown in Fig. 14.3. It clearly shows the LC phase and the inverse micellar (L_2) phase. This inverse micellar phase is relevant to the formation of water-in-oil emulsions. The interfacial tension between the micellar L_2 phase and water is about $1–2\,\mathrm{mN\,m^{-1}}$, and that between the L_2 and water is even lower. It is proposed that the L_2 phase forms an interfacial film during emulsification, and the droplet size distribution should then be expected to be related mainly to the interfacial and rheological properties of the L_2 phase.

The formation of a monomolecular film of the emulsifier at the oil/water (O/W) interface is a crucial factor in the emulsification process. Experimentally, it is easier to study lipid monolayers at the air/water (A/W) interface, compared to the O/W in-

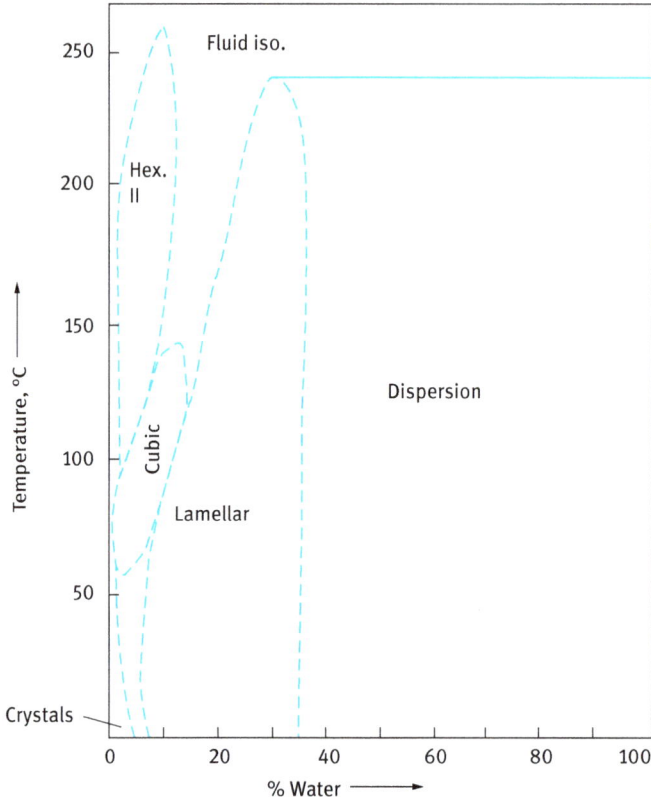

Fig. 14.2: Binary phase diagram of soybean lecithin-water system.

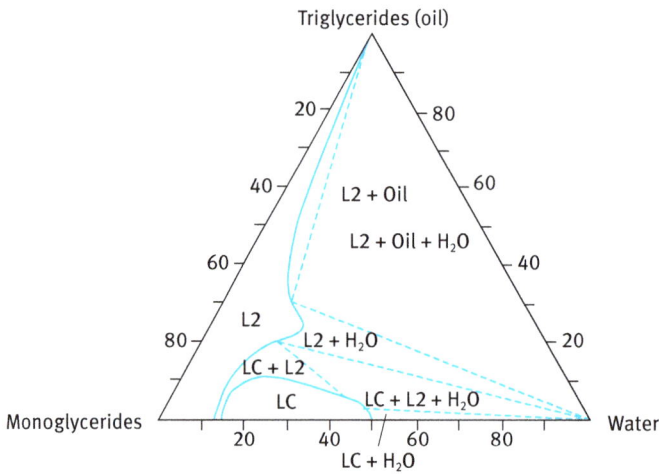

Fig. 14.3: Ternary phase diagram of soybean oil-sunflower oil monoglyceride and water.

terface. However, recently studies at the O/W interface became possible using drop profile techniques. The results showed similar trends to those observed at the A/W interface.

The results show a steep rise of surface pressure at a critical surfactant concentration ($\sim 10^{-6}$ mol dm^{-3}), and this concentration corresponds to the highest monomer concentration in bulk solution. Above 10^{-6} mol dm^{-3} the lipid monomers begin to associate. Surface pressure measurements at the air/water interface showed that the lipid molecules begin to associate to form a cubic structure. Monoglycerides of saturated fatty acids associate to form lamellar liquid crystalline phase or the gel phase at low concentrations. These condensed layers form at the oil-water interface at and above the critical temperature T_c, which is the temperature used for emulsification. These liquid crystalline phases play a major role in emulsion stabilisation. It is necessary to have enough polar lipids to form a stabilising film at the O/W interface, and this is not possible until the maximum of monomer concentration in the bulk is exceeded.

Hydrophilic emulsifiers, which in bulk form micellar solutions, exhibit a different monolayer behaviour. The interfacial tension γ shows a linear decrease with log concentration until the critical micelle concentration is reached, after which γ remains virtually constant with further increase in surfactant concentration. However, most surfactants used in food emulsions do not form micellar solutions.

A maximum in emulsion stability is obtained when three phases exist in equilibrium, and it was therefore proposed that the lamellar liquid crystalline phase stabilises the emulsion by forming a film at the O/W interface. This film provides a barrier against coalescence. This is illustrated in Fig. 14.4, which shows that the lamellar liquid crystalline phase exhibits a hydrophobic surface towards the oil and a hydrophilic surface towards the water. These multi-layers cause a significant reduction in the attraction potential, and they also produce a viscoelastic film with much higher viscosity than that of the oil droplet. In other words, the multi-layers produce a form of "mechanical barrier" against coalescence.

The rheological properties of monolayers of binary surfactant mixtures has been related to emulsion stability and to the structural properties of the lamellar liquid crystalline phases formed by the surfactant in water. It was suggested that the emulsi-

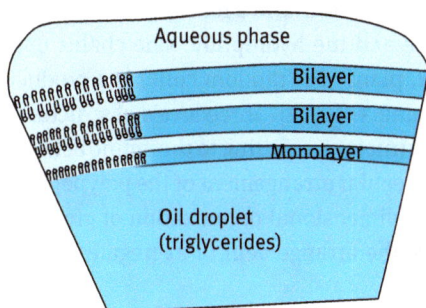

Fig. 14.4: Schematic representation of the lamellar liquid crystalline structure at the oil/water interface.

fier molecules adsorbed at the O/W interface will adopt the same hydrocarbon chain structure as they have in the bi-molecular lipid layer of lamellar mesophases. Liquid crystalline phases other than the lamellar phase can also occur at the O/W interface. Fluctuations of interfacial tension at the dodecane/water interface with a sodium sulphonate surfactant have been observed, and this was attributed to the formation of a liquid crystalline phase of the hexagonal type at the interface. The enhanced emulsion stability in the presence of lipid multi-layers at the interface is related to the reduced attraction potential. In addition, the multi-layer viscosity is considerably higher than that of the oil phase. Also the lamellar liquid crystalline phase results in a repulsive force, which is usually referred to as the hydration force.

Another important repulsive force that occurs when the surfactant film contains charged molecules, e.g. on addition of sodium stearate to lecithin, is the double layer repulsion arising from these ionogenic groups. An obvious consequence of the presence of charged chains is the increased distance between the surfactant bilayers.

The transition between the lamellar liquid crystalline phase and the gel phase can be utilised to stabilise the emulsion, provided the actual gel phase is stable. If an aqueous dispersion of the emulsifier is first formed, and the emulsification is then performed under cooling, an emulsion is formed, with the gel phase forming the O/W interface. Such an emulsion has a much higher stability when compared with that produced with the lamellar phase at the O/W interface. This is probably due to the higher mechanical stability of the crystalline lipid bilayers compared to bilayers with liquid chain conformation.

Another repulsive force between the lipid bi-layers in water is the hydration force, which is a short range force with exponential fall-off. This is related to repulsion between dipoles and induced dipoles. It is quite obvious that the hydration force will tend to inhibit coalescence of emulsion droplets with a multi-layer structure, as schematically shown in Fig. 14.12.

Proteins are also used as emulsifiers in many food emulsions. A protein is a linear chain of amino acids which assumes a three-dimensional shape dictated by the primary sequence of the amino acids in the chain. The side chains of the amino acids play an important role in directing the way in which the protein folds in solution. The hydrophobic (non-polar) side chains avoid interaction with water, while the hydrophilic (polar) side chains seek such interaction. This results in a folded globular structure, with the hydrophobic side chains inside and the hydrophilic side chains outside [9]. The final shape of the protein (helix, planar or "random coil") is a product of many interactions which form a delicate balance [10, 11]. Three levels of structural organisation have been suggested: (i) primary structure, referring to the amino acid sequence; (ii) secondary structure, denoting the regular arrangement of the polypeptide back bone; (iii) tertiary structure, as the three-dimensional organization of globular proteins. A quaternary structure consisting of the arrangement of aggregates of the globular proteins may also be distinguished.

Since proteins are used as emulsifying agents for oil-in-water emulsions, it is important to understand their interfacial properties, in particular the structural change that may occur on adsorption. The properties of protein adsorption layers differ significantly from those of simple surfactant molecules. In the first place, surface denaturation of the protein molecule may take place, resulting in unfolding of the molecule, at least at low surface pressures. Secondly, the partial molar surface area of proteins is large and can vary depending on the conditions for adsorption. The number of configurations of the protein molecule at the interface exceeds that in bulk solution, resulting in a significant increase of the non-ideality of the surface entropy. Thus, one cannot apply thermodynamic analysis, e.g. Langmuir adsorption isotherm, for protein adsorption. The question of reversibility vs irreversibility of protein adsorption at the liquid interface is still subject to a great deal of controversy. For that reason protein adsorption is usually described using statistical mechanical models. Scaling theories proposed by de Gennes [12] could also be applied.

One of the most important investigations of protein surface layers is to measure their interfacial rheological properties (e.g. its viscoelastic behaviour). Several techniques can be applied to study the rheological properties of protein layers, e.g. using constant stress (creep) or stress relaxation measurements. At very low protein concentrations, the interfacial layer exhibits Newtonian behaviour, independent of pH and ionic strength. At higher protein concentrations, the extent of surface coverage increases, and the interfacial layers exhibit viscoelastic behaviour, revealing features of solid-like phases. Above a critical protein concentration, protein-protein interactions become significant, resulting in a "two-dimensional" structure formation. The dynamics of formation of protein layers at the liquid-liquid interface should be considered in detail when one applies the protein molecules as stabilisers for emulsions. Several kinetic processes must be considered: solubilization of non-polar molecules resulting in the formation of associates in the aqueous phase; diffusion of solutes from bulk solution to the interface; adsorption of the molecules at the interface; orientation of the molecules at the liquid-liquid interface; formation of aggregation structures, etc.

When a protein is used as an emulsifier, it may adopt various conformations, depending on the interaction forces involved. The protein may adopt a folded or unfolded conformation at the oil/water interface. In addition, the protein molecule may interpenetrate in the lipid phase to various degrees. Several layers of proteins may also exist. The protein molecule may bridge one drop interface to another. The actual structure of the protein interfacial layer may be complex, combining any or all of the above possibilities. For these reasons, measurement of protein conformations at various interfaces still remains a difficult task, even when using several techniques such as UV, IR and NMR spectroscopy, as well as circular dichroism [13]. At an oil/water interface, the assumption is usually made that the protein molecule undergoes some unfolding, and this accounts for the lowering of the interfacial tension on protein adsorption. As mentioned above, multi-layers of protein molecules may be produced, and one should take into account the intermolecular interactions as well as the interaction with the lipid (oil) phase. Proteins act in a similar way to polymeric stabilisers (steric stabilization). However, the molecules with compact structures may precipitate to form small particles which accumulate at the oil/water interface. These particles stabilise the emulsions (sometimes referred to as Pickering emulsions) by a different mechanism. As a result of the partial wetting of the particles by the water and the oil, they remain at the interface. The equilibrium location at the interface provides the stability, since their displacement into the dispersed phase (during coalescence) results in an increase in the wetting energy.

From the above discussion, it is clear that proteins act as stabilisers for emulsions by different mechanisms, depending on their state at the interface. If the protein molecules unfold and form loops and tails, they provide stabilisation in a similar way to synthetic macromolecules. On the other hand, if the protein molecules form globular structures, they may provide a mechanical barrier that prevents coalescence. Finally, precipitated protein particles that are located at the oil/water interface provide stability as a result of the unfavourable increase in the wetting energy on their displacement. It is clear that in all cases, the rheological behaviour of the film plays an important role in the stability of the emulsions.

Proteins and polysaccharides are present in nearly all food emulsions [14]. The proteins are used as emulsion and foam stabilisers, whereas the polysaccharide acts as a thickener and also for water-holding. Both proteins and polysaccharides contribute to the structural and textural characteristics of many food emulsions through their aggregation and gelation behaviour. Several interactions between proteins and polysaccharides may be distinguished, ranging from repulsive to attractive interactions. The repulsive interactions may arise from excluded volume effects and/or electrostatic interaction. These repulsive interactions tend to be weak except at very low ionic strength (expanded double layers) or with anionic polysaccharides at pH values above the isoelectric point of the protein (negatively charged molecules). The attractive interaction can be weak or strong, and either specific or non-specific. A co-valent linkage between protein and polysaccharide represents a specific strong interaction. A non-specific protein-polysaccharide interaction may occur as a result of ionic, dipolar, hydrophobic or hydrogen bonding interaction between groups on the biopolymers. Strong attractive interaction may occur between positively charged protein (at a pH below its isoelectric point) and an anionic polysaccharide. In any particular system, the protein-polysaccharide interaction may change from repulsive to attractive as the temperature or solvent conditions (e.g. pH and ionic strength) change.

Aqueous solutions of proteins and polysaccharides may exhibit phase separation at finite concentrations. Two types of behaviour may be recognised, namely co-acervation and incompatibility. Complex co-acervation involves spontaneous separation into solvent-rich and solvent-depleted phases. The latter contains the protein-polysaccharide complex, which is caused by non-specific attractive protein-polysaccharide interaction, e.g. opposite charge interaction. Incompatibility is caused by spontaneous separation into two solvent-rich phases, one composed of predominantly protein and the other predominantly polysaccharide. Depending on the interactions, a gel formed from a mixture of two biopolymers may contain a coupled network, an interpenetrating network or a phase separated network. In food emulsions the two most important proteinaceous gelling systems are gelatin and casein micelles. An example of a covalent protein-polysaccharide interaction is that which is produced when gelatin reacts with propylene glycol alginate under mildly alkaline conditions. Non-covalent non-specific interaction occurs in mixed gels of gelatin with sodium alginate or low-methoxy pectin. In food emulsions containing protein and polysaccharide, any of the mentioned interactions may take place in the aqueous phase of the system. This results in specific structures with desirable rheological characteristics and enhanced stability. The nature of the protein-polysaccharide interaction affects the surface behaviour of the biopolymers and the aggregation properties of the dispersed droplets.

Weak protein-polysaccharide interactions may be exemplified by a mixture of milk protein (sodium casinate) and a hydrocolloid such as xanthan gum. Sodium casinate acts as the emulsifier and xanthan gum (with a molecular weight in the region of $2 \cdot 10^6$ Da) is widely used as a thickening agent and a synergistic gelling agent (with locust bean gum). In solution, xanthan gum exhibits pseudo-plastic behaviour which is maintained over a wide range of temperature, pH and ionic strength. Xanthan gum at concentrations exceeding 0.1 % inhibits creaming of emulsion droplets by producing a gel-like network with a high residual viscosity. At lower xanthan gum concentrations (< 0.1 %) creaming is enhanced as a result of depletion flocculation. Other hydrocolloids such as carboxymethylcellulose (with a lower molecular weight than xanthan gum) are less effective in reducing creaming of emulsions.

Covalent protein-polysaccharide conjugates are sometimes used to avoid any flocculation and phase separation which is produced with weak non-specific protein-polysaccharide interactions. An example of such conjugates is that produced with globulin-dextran or bovine serum albumin-dextran. These conjugates produce emulsions with smaller droplets and narrower size distribution, and they stabilise the emulsion against creaming and coalescence.

One of the most important aspects of polymer-surfactant systems is their ability to control stability and rheology over a wide range of composition [14]. Surfactant molecules which bind to a polymer chain generally do so in clusters that closely resemble the micelles formed in the absence of polymer [15]. If the polymer is less polar or contains hydrophobic regions or sites, there is an intimate contact between the micelles and polymer chain. In such a situation, the contact between one surfactant micelle and two polymer segments will be favourable. The two segments can be in the same polymer chain or in two different chains, depending on the polymer concentration. For a dilute solution, the two segments can be in the same polymer chain, whereas in more concentrated solutions the two segments can be in two polymer chains with significant chain overlap. The cross-linking of two or more polymer chains can lead to network formation and dramatic rheological effects.

Surfactant-polymer interaction can be treated in different ways, depending on the nature of the polymer. A useful approach is to consider the binding of surfactant to a polymer chain as a co-operative process. The onset of binding is well defined and can be characterised by a critical association concentration (CAC). The latter decreases with increase in the alkyl chain length of the surfactant. This implies an effect of polymer on surfactant micellisation. The polymer is considered to stabilise the micelle by short- or long-range electrostatic interaction. The main driving force for surfactant self-assembly in polymer-surfactant mixtures is generally the hydrophobic interaction

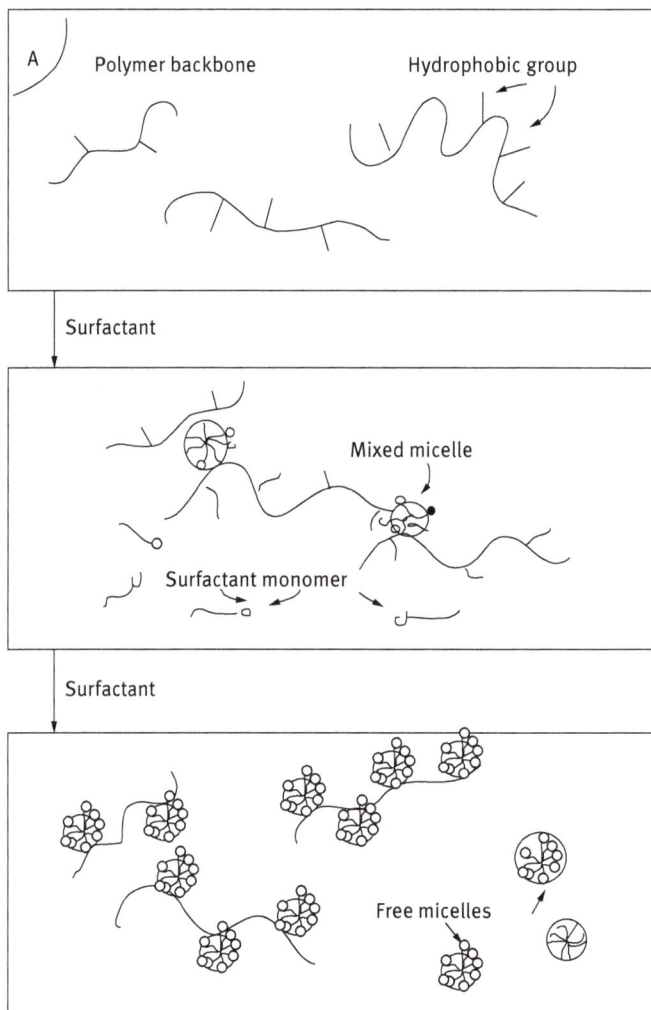

Fig. 14.5: Schematic representation of the interaction between hydrophobically modified polymer chains and surfactant micelles.

between the alkyl chains of the surfactant molecules. Ionic surfactants often interact significantly with both non-ionic and ionic polymers. This can be attributed to the unfavourable contribution to the energetics of micelle formation from the electrostatic effects and their partial elimination due to charge neutralization or lowering of the charge density. For non-ionic surfactants, there is little to gain in forming micelles in the presence of a polymer, and hence the interaction between non-ionic surfactants and polymers is relatively weak. However, if the polymer chain contains hydrophobic segments or groups, e.g. with block copolymers, the hydrophobic polymer-surfactant interaction will be significant.

For hydrophobically modified polymers (such as hydrophobically modified hydroxyethyl cellulose or polyethylene oxide), the interaction between the surfactant micelles and the hydrophobic chains on the polymer can result in the formation of cross links, i.e. gel formation. This is schematically represented in Fig. 14.5. However, at high surfactant concentrations, there will be more micelles which can interact with the individual polymer , and the cross links are broken.

The above interactions are manifested in the variation of viscosity with surfactant concentration. Initially, the viscosity shows an increase with an increase in surfactant concentration, reaching a maximum, and then decreases with a further increase in surfactant concentration. The maximum is consistent with the formation of crosslinks, and the decrease after that indicates destruction these cross links (see Fig. 14.5).

14.2.2 Surfactant association structures, micro-emulsions and emulsions in food [3]

A typical phase diagram of a ternary system of water, ionic surfactant and long chain alcohol (co-surfactant) is shown in Fig. 14.6. The aqueous micellar solution A solubilises some alcohol (spherical normal micelles), whereas the alcohol solution dissolves huge amounts of water forming inverse micelles, B. These two phases are not in equilibrium with each other, but are separated by a third region, namely the lameller liquid crystalline phase. These lamellar structures and their equilibrium with the aqueous micellar solution (A) and the inverse micellar solution (B) are the essential elements for both microemulsion and emulsion stability [3].

Micro-emulsions are thermodynamically stable, and they form spontaneously (primary droplets a few nms in size), whereas emulsions are not thermodynamically stable, since the interfacial free energy is positive and dominant in the total free energy. This difference can be related to the difference in bending energy between the two systems [3]. With micro-emulsions, which contain very small droplets, the bending energy (negative contribution) is comparable to the stretching energy (positive contribution) and hence the total surface free energy is extremely small ($\sim 10^{-3}$ mN m^{-1}). With macroemulsions, on the other hand, the bending energy is negligible (small curvature of the large emulsion drops), and hence the stretching energy dominates

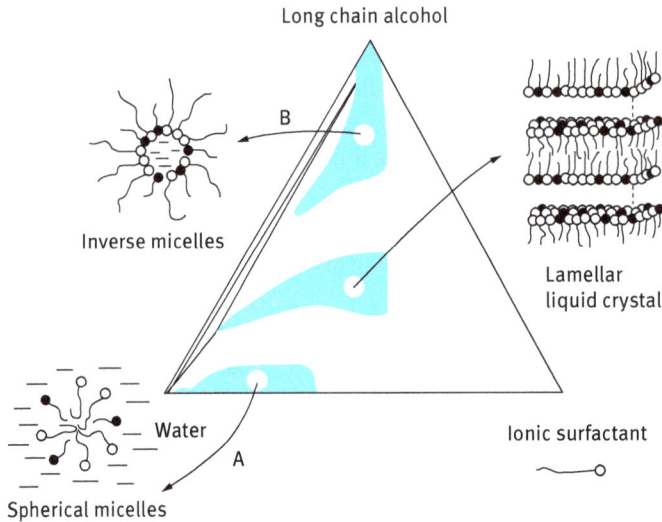

Fig. 14.6: Ternary-phase diagram of water, an anionic surfactant and long chain alcohol (co-surfactant).

the total surface free energy, which is now large and positive (a few $mN\,m^{-1}$). The micro-emulsion may be related to the micellar solutions A and B shown in Fig. 14.6. A W/O micro-emulsion is obtained by adding a hydrocarbon to the inverse micellar solution B, whereas an O/W micro-emulsion emanates from the aqueous micellar solution A. These micro-emulsion regions are in equilibrium with the lamellar liquid crystalline structure. To maximise the micro-emulsion region, the lamellar phase has to be destabilised, as for example by the addition of a relatively short chain alcohol such as pentanol. In contrast for a macro-emulsion, with its large radius, the parallel packing of the surfactant/cosurfactant is optimal, and hence the co-surfactant should be of a chain length similar to that of the surfactant.

From the above discussion, it is clear that a surfactant/cosurfactant combination for a micro-emulsion is of little use for stabilising an emulsion. This is a disadvantage when a multiple emulsion of the W/O/W type is to be formulated, whereby the W/O system is a micro-emulsion. This problem was resolved by Larsson [13], who used a surfactant combination to stabilise the micro-emulsion and a polymer to stabilise the emulsion.

The formulation of food systems as micro-emulsions is not easy, since the addition of triglycerides to inverse micellar systems results in a phase change to a lamellar liquid crystalline phase. The latter has to be destabilised by means other than adding co-surfactants, which are normally toxic. An alternative approach to destabilising the lamellar phase is to use a hydrotrope, a number of which are allowed in food products.

As discussed above, for emulsion stabilisation in food systems lamellar liquid crystalline structures are ideal. At the interface, the liquid crystals serve as a viscous

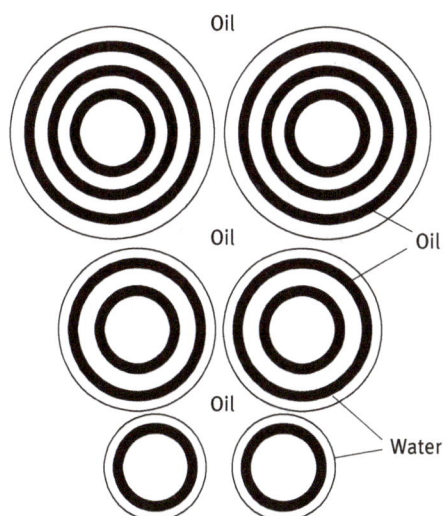

Fig. 14.7: Schematic representation of emulsions containing liquid crystalline structures.

barrier to accept and dissipate the energy of flocculation [16]. This is illustrated in Fig. 14.7, which shows the coalescence process of a droplet covered with a lamellar liquid crystal. It consists of two stages; at first the layers of the liquid crystals are removed two by two, and the terminal step is the disruption of the final bilayer of the structure.

The initiation of the flocculation process leads to very small energy changes, and good stability is assumed as long as the liquid crystal remains adsorbed. This adsorption is the result of its structure. At the interface, the final layer towards the aqueous phase terminates with the polar group, while the layer towards the oil finishes with the methyl layer. In this manner, the interfacial free energy is a minimum.

14.3 Emulsions in cosmetics and personal care formulations

Several cosmetic formulations can be identified: lotions, hand creams (cosmetic emulsions), nanoemulsions, multiple emulsions, liposomes, shampoos and hair conditioners, sunscreens and colour cosmetics. The ingredients used must be safe and should not cause any damage to the organs with which they come in contact.

Cosmetic and toiletry products are generally designed to deliver a function benefit and to enhance the psychological well-being of consumers by increasing their aesthetic appeal. Thus, many cosmetic formulations are used to clean hair, skin, etc. and impart a pleasant odour, make the skin feel smooth and provide moisturising agents, provide protection against sunburn etc. In many cases, cosmetic formulations are designed to provide a protective, occlusive surface layer which either prevents the penetration of unwanted foreign matter or moderates the loss of water from the skin

[17, 18]. In order to have consumer appeal, cosmetic formulations must meet stringent aesthetic standards such as texture, consistency, pleasing colour and fragrance, convenience of application, etc. This results in most cases in complex systems consisting of several components of oil, water, surfactants, colouring agents, fragrants, preservatives, vitamins, etc. In recent years, there has been considerable effort in introducing novel cosmetic formulations that provide great beneficial effects to the customer, such as sunscreens, liposomes and other ingredients that may keep healthy skin and provide protection against drying, irritation, etc. Since cosmetic products come in close contact with various organs and tissues of the human body, a most important consideration for choosing ingredients to be used in these formulations is their medical safety. Many of the cosmetic preparations are left on the skin after application for indefinite periods of time and, therefore, the ingredients used must not cause allergic reactions, sensitisation or irritation. The ingredients used must be free of impurities which have toxic effects.

One of the main areas of interest of cosmetic formulations is their interaction with the skin [19]. This is certainly an interfacial phenomenon involving wetting, spreading and adhesion. The top layer of the skin, which is the main barrier against water loss, is the stratum corneum, which protects the body from chemical and biological attack [20]. This layer is very thin, approximately 30 μm, and it consists of ~ 10 % by weight of lipids organised into bi-layer structures (liquid crystalline) which at high water content is soft and transparent [21, 22].

To achieve the above criteria, several systems are formulated as emulsions, e.g.: oil-in-water (O/W) emulsions, water-in-oil (W/O) emulsions, nano-emulsions and multiple emulsions. Most of these emulsions contain "self-assembly" structures, e.g. micelles (spherical, rod-shaped, lamellar), liquid crystalline Phases (hexagonal, cubic or lamellar), liposomes (multilamellar bilayers) or vesicles (single bi-layers). They also contain "thickeners" (polymers or particulate dispersions) to control their rheology.

As mentioned above, surfactants used in cosmetic formulations must be completely free of allergenics, sensitisers and irritants. To minimise medical risks, cosmetic formulators tend to use polymeric surfactants, which are less likely to penetrate beyond the stratum corneum and hence are less likely to cause any damage.

Conventional surfactants of the anionic, cationic, amphoteric and non-ionic types are used in cosmetic systems. Besides the synthetic surfactants used in preparation of cosmetic systems such as emulsions, creams, suspensions, etc., several other naturally occurring materials have been introduced, and there has been a trend in recent years to use such natural products more widely, in the belief that they are safer for application. Polymeric surfactants of the A–B, A–B–A and BA_n are also used in many cosmetic formulations. The macro-molecular surfactants possess considerable advantages for use in cosmetic ingredients. The most commonly used materials are the ABA block copolymers, with A being poly(ethylene oxide) and B poly(propylene oxide) (pluronics). On the whole, polymeric surfactants have much lower toxicity, sensitisation and irritation potentials, provided they are not contaminated with traces of the parent monomers. As will be discussed in the section on emulsions, these molecules provide greater stability, and in some case they can be used to adjust the viscosity of the cosmetic formulation.

In recent years, there has been a great trend towards using silicone oils for many cosmetic formulations. In particular, volatile silicone oils have found application in many cosmetic emulsions, owing to the pleasant dry sensation they impart to the skin. These volatile silicones evaporate without unpleasant cooling effects or without leaving a residue. Due to their low surface energy, silicone helps spread the various active ingredients over the surface of hair and skin. The chemical structure of the silicone compounds used in cosmetic preparations varies according to the application. The backbones can carry various attached "functional" groups, e.g. carboxyl, amine, sulfhydryl, etc. [23]. While most silicone oils can be emulsified using conventional hydrocarbon surfactants, there has been a trend in recent years to use silicone surfactants to produce emulsions [24].

As mentioned above, cosmetic emulsions need to contain a number of benefits. For example, such systems should offer a functional benefit such as cleaning (e.g. hair, skin, etc.), provide a protective barrier against water loss from the skin, and in some cases they should screen out damaging UV light (in which case a sunscreen agent such as titania is incorporated into the emulsion). These systems should also impart a pleasant odour and make the skin feel smooth. Emulsions, both oil-in-water (O/W) and water-in-oil (W/O), are used in cosmetic applications. The main physicochemical characteristics that need to be controlled in cosmetic emulsions are their formation and stability on storage as well as their rheology, which controls spreadability and skin feel. The life span of most cosmetic and toiletry brands is relatively short (3–5 years), and hence development of the product needs to be quick. For this reason, accelerated storage testing is needed for prediction of stability and change of rheology with time. These accelerated tests represent a challenge to the formulation chemist.

As mentioned above, the main criterion for any cosmetic ingredient should be medical safety (free of allergenics, sensitisers, irritants and impurities which have systemic toxic effects). These ingredients should be suitable for producing stable emulsions that can deliver the functional benefit and the aesthetic characteristics. The main composition of an emulsion is the water and oil phases and the emulsifier. Several water soluble ingredients may be incorporated in the aqueous phase and oil soluble ingredients in the oil phase. Thus, the water phase may contain functional materials such as proteins, vitamins, minerals and many natural or synthetic water-soluble polymers. The oil phase may contain perfumes and/or pigments (e.g. in make-up). The oil phase may be a mixture of several mineral or vegetable oils. Examples of oils used in cosmetic emulsions are linolin and its derivatives, paraffin and silicone oils. The oil phase provides a barrier against water loss from the skin.

The process of emulsion formation is determined by the property of the interface, in particular the interfacial tension which is determined by the concentration and type of the emulsifier. This was described in detail in Chapter 2. Several emulsifiers, mostly non-ionic or polymeric, are used for preparation of O/W or W/O emulsions and their subsequent stabilisation. For W/O emulsion, the hydrophilic-lipophilic balance (HLB) range of the emulsifier is in the range of 3–6, whereas for O/W emulsions this range is 8–18, as described in Chapter 7.

Cosmetic emulsions are usually referred to as skin creams, which may be classified according to their functional application. For the manufacture of cosmetic emulsions, it is necessary to control the process which determines the droplet size distribution, since this controls the rheology of the resulting emulsion. Usually, one starts to make the emulsion on a lab scale (of the order of 1–2 litres), which has to be scaled-up to a pilot plant and manufacturing scale. At each stage, it is necessary to control the various process parameters which need to be optimised to produce the desirable effect. It is necessary to relate the process variable from the lab to the pilot plant to the manufacturing scale, and this requires a great deal of understanding of emulsion formation controlled by the interfacial properties of the surfactant film. Two main factors should be considered, namely the mixing conditions and selection of production equipment. For proper mixing, sufficient agitation, for producing turbulent flow is necessary in order to break up the liquid (disperse phase) into small droplets. Various parameters should be controlled, such as flow rate and turbulence, type of impellers, viscosity of the internal and external phases and the interfacial properties such as surface tension, surface elasticity and viscosity. The selection of production equipment depends on the characteristics of the emulsion to be produced. Propeller and turbine agitators are normally used for low and medium viscosity emulsions. Agitators capable of scraping the walls of the vessel are essential for high viscosity emulsions. Very high shear rates can be produced by using ultrasonics, colloid mills and homogenisers. It is essential to avoid too much heating in the emulsion during preparation, which may have undesirable effects such as flocculation and coalescence.

The rheological properties of a cosmetic emulsion which need to be achieved depend on the consumer perspective, which is very subjective. However, the efficacy and aesthetic qualities of a cosmetic emulsion are affected by their rheology. For example, with moisturising creams one requires fast dispersion and deposition of a continuous protective oil film over the skin surface. This requires a shear thinning system (see below).

For characterisation of the rheology of a cosmetic emulsion, one needs to combine several techniques, namely steady state, dynamic (oscillatory) and constant stress (creep) measurements [25–27]. A brief description of these techniques is given below.

In steady-state measurements one measures the shear stress (τ)–shear rate (γ) relationship using a rotational viscometer. A concentric cylinder or cone and plate geometry may be used, depending on the emulsion consistency. Most cosmetic emulsions are non-Newtonian, usually pseudo-plastic, as illustrated in Fig. 14.11. In this case the viscosity decreases with applied shear rate (shear thinning behaviour (Fig. 14.8), but

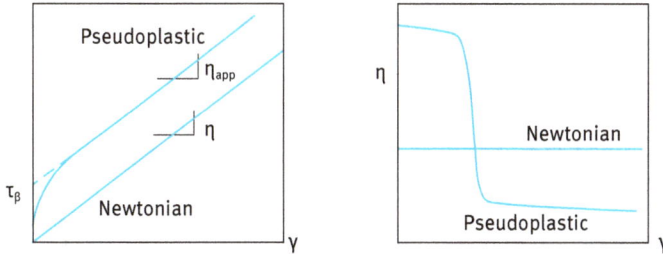

Fig. 14.8: Schematic representation of Newtonian and non-Newtonian (pseudo-plastic) flow.

at very low shear rates the viscosity reaches a high limiting value (usually referred to as the residual or zero shear viscosity).

For the above pseudo-plastic flow, one may apply a power law fluid model, a Bingham model [27] or a Casson model [28]. These models are represented by the following equations, respectively:

$$\tau = \eta_{app}\gamma^n \tag{14.1}$$

$$\tau = \tau_\beta + \eta_{app}\gamma \tag{14.2}$$

$$\tau^{1/2} = \tau_c^{1/2} + \eta_c^{1/2}\gamma^{1/2}, \tag{14.3}$$

where n is the power in shear rate, which is less than 1 for a shear thinning system (n is sometimes referred to as the consistency index), τ_β is the Bingham (extrapolated) yield value, η is the slope of the linear portion of the τ–γ curve, usually referred to as the plastic or apparent viscosity, τ_c is the Casson's yield value and η_c is the Casson's viscosity.

In dynamic (oscillator) measurements, a sinusoidal strain, with frequency v in Hz or ω in rad s^{-1} ($\omega = 2\pi v$) is applied to the cup (of a concentric cylinder) or plate (of a cone and plate), and the stress is measured simultaneously on the bob or the cone connected to a torque bar. The angular displacement of the cup or the plate is measured using a transducer. For a viscoelastic system, such as the case with a cosmetic emulsion, the stress oscillates with the same frequency as the strain, but out of phase [23]. This is illustrated in Fig. 14.9, which shows the stress and strain sine waves for a viscoelastic system. From the time shift between the sine waves of the stress and strain, Δt, the phase angle shift δ is calculated

$$\delta = \Delta t\omega. \tag{14.4}$$

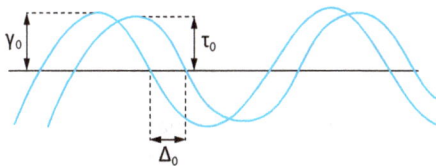

Fig. 14.9: Schematic representation of stress and strain sine waves for a viscoelastic system.

The complex modulus, G^*, is calculated from the stress and strain amplitudes (τ_0 and γ_0 respectively), i.e.

$$G^* = \frac{\tau_0}{\gamma_0}. \tag{14.5}$$

The storage modulus, G', which is a measure of the elastic component, is given by the following expression:

$$G' = |G^*| \cos \delta \tag{14.6}$$

The loss modulus, G'', which is a measure of the viscous component, is given by the following expression:

$$G'' = |G^*| \sin \delta, \tag{14.7}$$

and

$$|G^*| = G' + iG'', \tag{14.8}$$

where i is equal to $(-1)^{1/2}$.

The dynamic viscosity, η', is given by the following expression:

$$\eta' = \frac{G''}{\omega}. \tag{14.9}$$

In dynamic measurements one carries two separate experiments. Firstly, the viscoelastic parameters are measured as a function of strain amplitude, at constant frequency, in order to establish the linear viscoelastic region, where G^*, G' and G'' are independent of the strain amplitude. This is illustrated in Fig. 14.10, which shows the variation of G^*, G' and G'' with γ_0. It can be seen that the viscoelastic parameters remain constant up to a critical strain value, γ_{cr}, above which G^* and G' start to decrease and G'' starts to increase with further increase in the strain amplitude. Most cosmetic emulsions produce a linear viscoelastic response up to appreciable strains ($> 10\,\%$), indicative of structure build-up in the system ("gel" formation). If the system shows a short linear region (i.e., a low γ_{cr}), it indicates lack of a "coherent" gel structure (in many cases this is indicative of strong flocculation in the system).

Once the linear viscoelastic region is established, measurements are then made of the viscoelastic parameters, at strain amplitudes within the linear region, as a function of frequency. This is schematically illustrated in Fig. 14.11, which shows the variation of G^*, G' and G'' with v or ω. It can be seen that below a characteristic frequency,

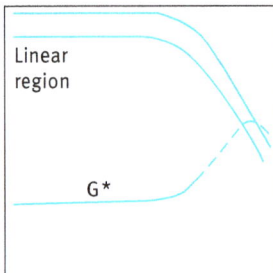

Fig. 14.10: Schematic representation of the variation of G^*, G' and G'' with strain amplitude (at a fixed frequency).

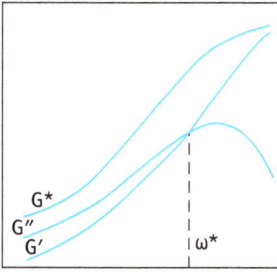

Fig. 14.11: Schematic representation of the variation of G^*, G' and G'' with ω for a viscoelastic system.

v^* or ω^*, $G'' > G'$. In this low frequency regime (long time scale), the system can dissipate energy as viscous flow. Above v^* or ω^*, $G' > G''$, since in this high frequency regime (short time scale) the system is able to store energy elastically. Indeed at sufficiently high frequency G'' tends to zero and G' approaches G^* closely, showing little dependency on frequency. The relaxation time of the system can be calculated from the characteristic frequency (the cross-over point) at which $G' = G''$, i.e.

$$t^* = \frac{1}{\omega^*} \, . \tag{14.10}$$

Many cosmetic emulsions behave as semi-solids with long t^*. They only show an elastic response within the practical range of the instrument, i.e. $G' \gg G''$, and it exhibits small dependence on frequency. Thus, the behaviour of many emulsions creams is similar to many elastic gels. This is not surprising, since in most of cosmetic emulsions systems, the volume fraction of the disperse phase of most cosmetic emulsions is fairly high (usually > 0.5), and in many systems a polymeric thickener is added to the continuous phase for stabilisation of the emulsion against creaming (or sedimentation) and to produce the right consistency for application.

In creep (constant stress) measurements [25], a stress τ is applied on the system, and the deformation γ or the compliance $J = \gamma/\tau$ is followed as a function of time. A typical example of a creep curve is shown in Fig. 14.12. At $t = 0$, i.e. just after the application of the stress, the system shows a rapid elastic response characterized by an instantaneous compliance J_o which is proportional to the instantaneous modulus G_o. Clearly at $t = 0$, all the energy is stored elastically in the system. At $t > 0$, the compliance shows a slow increase, since bonds are broken and reformed but at different rates. This retarded response is the mixed viscoelastic region. At sufficiently large time scales, which depend on the system, a steady state may be reached with a constant shear rate. In this region, J shows a linear increase with time, and the slope of the straight line gives the viscosity, $\eta\tau$, at the applied stress. If the stress is removed, after the steady state is reached J decreases, and the deformation reverses sign, but only the elastic part is recovered. By carrying out creep curves at various stresses (starting from very low values, depending on the instrument sensitivity), one can obtain the viscosity of the emulsion at various stresses. A plot of η_τ vs τ shows the typical behaviour shown in Fig. 14.13. Below a critical stress, τ_β, the system shows a Newtonian

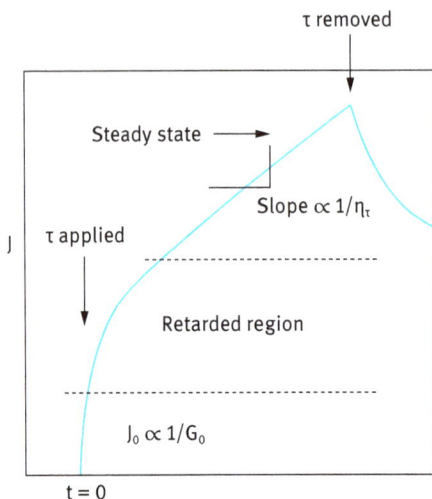

Fig. 14.12: Typical creep curve for a viscoelastic system.

region with a very high viscosity, usually referred to as the residual (or zero shear) viscosity. Above τ_β, the emulsion shows a shear thinning region, and ultimately another Newtonian region with a viscosity much lower than $\eta(0)$ is obtained. The residual viscosity gives information on the stability of the emulsion on storage. The higher the value of $\eta(0)$ the lower the creaming or sedimentation of the emulsion. The high stress viscosity gives information on the applicability of the emulsion, such as its spreading and film formation. The critical stress τ_β gives a measure of the true yield value of the system, which is an important parameter both for application purposes and the long-term physical stability of the cosmetic emulsion.

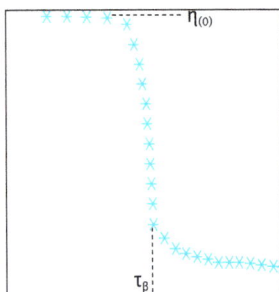

Fig. 14.13: Variation of viscosity with applied stress for a cosmetic emulsion.

It is clear from the above discussion that rheological measurements of cosmetic emulsions are very valuable in determining the long term physical stability of the system as well as its application. This subject has attracted considerable interest in recent years with many cosmetic manufacturers. Apart from its value in the above mentioned assessment, one of the most important considerations is to relate the rheological parameters to the consumer perception of the product. This requires careful measurement of the various rheological parameters for a number of cosmetic products and relating these parameters to the perception of expert panels who assess the consistency of the product, its skin feel, spreading, adhesion, etc. It is claimed that the rheological properties of an emulsion cream determine the final thickness of the oil layer, the moisturising efficiency and its aesthetic properties such as stickiness, stiffness and oiliness (texture profile). Psychophysical models may be applied to correlate rheology with consumer perception and a new branch of psychorheology may be introduced.

14.4 Emulsions in pharmacy

Several pharmaceutical products are formulated as emulsions [25]: (i) parenteral emulsion systems, e.g. parenteral nutritional emulsions, lipid emulsions as drug carriers; (ii) perfluorochemical emulsions as artificial blood substitute; (iii) emulsions as vehicles for vaccines; (iv) topical Formulations, e.g. for treatment of some skin diseases (dermatitis). For preparation of any of the above systems, the formulation chemist must choose an optimum emulsifier system that is suitable for the preparation of the emulsion and maintenance of its long term physical stability. The oil that may be used as a drug carrier has to be non-toxic, e.g. vegetable oils (soybean and safflower), synthetic glycerides (including simulated human fats), acetoglycerides. The emulsifier system chosen should be safe (having no undesirable toxic effects) and it should be approved by the FDA (Food and Drug Administration). The following are examples of some approved emulsifiers:
(i)   anionic: sodium cholate – bile salts;
(ii) zwitterionic: lecithin (mainly phosphatidyl choline, phosphatidylethanolamine);
(iii) non-ionic: polyethylene glycol stearate – polyoxyethylene mononstearate (myrj), Sorbitan esters (spans) and their ethoxylates (tweens), poloxamers (polyoxyethylene – polyoxypropylene block copolymers).

For non-ionic surfactants, particularly those of the ethoxylate type, a selection can be made based on the hydrophilic-lipophilic-balance (HLB) concept that was described in Chapter 7. A closely related system is based on the phase inversion temperature (PIT) concept, also described in detail in Chapter 7. For the preparation of macroemulsion (with droplet size in the range 1–5 μm), such as in many topical application creams, high-speed stirrers such as Silverson or Ultraturrax can be used. However, for parenteral emulsions (such as fat emulsions and anaesthetics), a much smaller particle size range is required, usually in the range 200–500 nm (sometimes referred to as nano-emulsions). These systems can be prepared using high temperature and/or high pressure techniques (using homogenisers such as the Microfluidiser).

The role of surfactant in emulsion formation is crucial, and this was described in detail in Chapter 5. It reduces the oil/water interfacial tension, γ_{OW} by adsorption at the interface. The droplet size R is directly proportional to γ_{OW}. It also enhances deformation and break-up of the droplets by reducing the Laplace pressure p:

$$p = \frac{2\gamma_{OW}}{R} . \tag{14.11}$$

It also prevents coalescence during emulsification by creating a tangential stress at the interface (Gibbs–Marangoni effect), as described in Chapter 5. During emulsification, deformation of the droplets results in increase in the interfacial area A, and hence the surfactant molecules will show a concentration gradient at the interface that gives an interfacial dilational (Gibbs) elasticity ε:

$$\varepsilon = \frac{d\gamma}{d\ln A} . \tag{14.12}$$

As a result of this interfacial tension gradient, surfactant molecules will diffuse to the regions with higher γ and they carry liquid with them (i.e. they force liquid in between the droplets), and this prevents coalescence (Marangoni effect).

Several breakdown processes of emulsions can be distinguished on standing, and these were described in Chapter 1. Apart from the process of creaming or sedimentation that is governed by gravity, all other breakdown processes are determined by the interaction forces (energies) between the droplets. These interaction energies were described in detail in Chapter 3.

Lipid emulsions are used for parenteral nutrition, e.g Intralipid that consists of 10–20 % soybean, 1.2 % egg lecithin and 2.5 % glycerol. The advantage of a fat emulsion is that a large amount of energy can be delivered in a small volume of isotonic fluid via a peripheral vein. The main problem with these fat emulsions is their long term stability. A wide range of non-ionic emulsifying agents have been investigated as potential emulsifying agents for intravenous fat. The commercial fat emulsions employed in parenteral nutrition are stabilised by egg lecithin. The egg lecithin is a complex mixture of phospholipids with the following composition: Phosphatidyl choline (PC) 7.3 %, Lysophosphatidyl choline (LPC) 5.8 %, Phosphatidyl ethanolamine 15.0 %, Lysophosphatidyl ethanolamine (LPE) 2.1 %, Phosphatidylinositol (PI) 0.6 %, Sphingomyelin (SP) 2.5 %.

In the development of suitable fat emulsions, pure PC and PE were employed, but the emulsion had poor stability. This is due to the lack of formation of an electrical or mechanical barrier against coalescence. Introduction of ionic lipids such as phosphatidic acid (PA) and phosphatidylserine was essential to improve the stability of the emulsion.

An essential requirement for fat emulsions is a small particle size (200–500 nm), and this required the use of high pressure homogenisers. It was essential to store the emulsion at various temperatures and investigate any increase in fatty acid composition which causes lipoprotein lipase reactions. Also increase in droplet size increased the toxicity of the emulsion. Addition of drugs and nutrients to fat emulsions can also cause instability and/or cracking of the emulsion. Following the administration of fat emulsions to the body, it will be distributed rapidly throughout the circulatory system and then cleared.

Another application of emulsions in pharmacy is the perfluorochemical emulsions that are used as blood substitutes. Perfluorochemicals have the ability to dissolve large quantities of oxygen and hence can be used as red blood substitutes. Several emulsifying agents have been examined, and the best stability was obtained using the block copolymers of polyethylene oxide polypropylene oxide (poloxamers or pluronics, e.g. pluronic F68). Several advantages of fluorochemical emulsions should be mentioned: good shelf life, good stability in surgical procedures, no blood-group incompatibility problems, ready accessibility, and no problem with hepatitis.

The final emulsion should satisfy the following criteria: (a) low toxicity; (b) no adverse interaction with normal blood; (c) little effect on blood clotting; (d) satisfactory oxygen and carbon dioxide exchange; (e) satisfactory rheological characteristics; (f) satisfactory clearance from the body.

To date the use of perflurocarbon emulsions as blood substitutes have been investigated using animal studies, although a product from Japan (Fluosol-DA) containing perfluorodecalin has also been studied in humans. The formulation of a suitable perfluorocarbon emulsion is still in its infancy, since oils that produce stable emulsions are not cleared from the body, and the choice of a suitable emulsifier is still difficult. The most suitable emulsifier system was based on a mixture of lecithin and poloxamer. The preferred oil, perfluorodecalin, initially had a fine droplet size, but on storage the droplets grew in size due to an Ostwald ripening mechanism. The size of the emulsion droplets can have a pronounced effect on the biological results. Fluosol-DA has a mean particle size in the region of 100–200 nm. Large particles were shown to have toxic effects.

14.5 Emulsions in agrochemicals

Many agrochemicals are formulated as oil-in-water (O/W) emulsion concentrates (EWs) [29]. These systems offer many advantages over the more traditionally used emulsifiable concentrates (ECs). By using an O/W system one can reduce the amount of oil in the formulation, since in most cases a small proportion of oil is added to the agrochemical oil (if this has a high viscosity) before emulsification. In some cases if the agrochemical oil has a low to medium viscosity, one can emulsify the active ingredient directly into water. With many agrochemicals with a ow melting point, which is not suitable for the preparation of a suspension concentrate, one can dissolve the active ingredient in a suitable oil, and the oil solution is then emulsified into water. EWs which are aqueous-based produce less hazard to the operator reducing any skin irritation. In addition, in most cases EWs are less phytotoxic to plants when compared with ECs. The O/W emulsion is convenient for incorporation of water soluble adjuvants (mostly surfactants). EWs can also be less expensive when compared to ECss since a lower surfactant concentration is used to produce the emulsion and also one replaces a great proportion of oil by water. The only drawback of EWs compared to ECs is the need to use high-speed stirrers and/or homogenisers to obtain the required droplet size distribution. In addition EWs require control and maintenance of its physical stability. As discussed in Chapter 2, EWs are only kinetically stable, and one has to control the breakdown process that occurs during storage, such as creaming or sedimentation, flocculation, Ostwald ripening, coalescence and phase inversion. The various methods which can be applied to reduce creaming/sedimentation, flocculation, Ostwald ripening, coalescence and phase inversion, have been described in detail in Chapters 8–12.

14.6 Rolling oil and lubricant emulsions

Rolling oil emulsions are used during the process of strip steel process, where the strip thickness is reduced by a factor of about 4 to a final thickness of between 0.45 and 3 mm [30]. The strip speeds are of the order of magnitude of 600 to 1000 m min^{-1}. Rolling oil emulsions must perform two main tasks: dissipating the heat and providing sufficient lubrication so that an even surface texture is imparted to the strip steel at such high speeds. The lubricant film on the steel surface would interfere with most of the subsequent processing steps, e.g. welding, electroplating or phosphatising. Rolling oil emulsions therefore contain only a low concentration of oil ranging from 1.5 % for the first rolling stand down to 0.5 % for the fourth stand. In view of the large steel throughput, the quantities of emulsion used are enormous. On average, between 4 and 5 m^3 min^{-1} are consumed per rolling stand, giving a consumption rate of 0.2 kg t^{-1} rolled steel. The lubricating emulsion passes through a cycle; fresh rolling oil emulsion is applied at the stand. Some of the water evaporates in the roll gap and the emulsion and the oil form a lubricating film on the surface of steel. The residual emulsion is run off. The run-off emulsion, whose composition now differs from that of the original emulsion, is purified, the water and oil content is relinquished and the emulsion is returned to the cycle. The main components of cold rolling lubricants are mineral oils and castor oil. Synthetic ester oils have proved superior to natural triglycerides on account of their better resistance to hydrolysis and their temperature stability. Mixtures of alkyl ethoxylates are frequently employed for emulsifying purposes, and in some cases protective colloids are also added to provide steric stabilisation. It is crucial for the rolling result that the emulsifier has the optimum composition. The development of controlled particle size emulsifiers brought a breakthrough in cold rolling technology by enabling stable emulsions with such large droplets and narrow size distribution to be produced in the rolling process.

The problems of heat dissipation and lubrication are common to many activities in the metal processing industry. Cooling lubricants are needed during drilling and cutting operations, for example to ensure that the heat is dissipated quickly enough to protect the work piece against thermal damage. In addition, the cooling lubricant stream carries metal chips away from the processing zone and covers the freshly exposed metal surface with a protective film of emulsion. The lubricant emulsions used are of the O/W type, with a more or less substantial proportion of oil. Both nanoemulsions and microemulsions are also used as lubricating systems.

References

[1]   N. J. Krog and T. H. Riisom, in: P. Becher (ed.), *Encyclopedia of Emulsion Technology*, vol. 2, pp. 321–365, Marcel Dekker, New York, 1985.

[2]   E. N. Jaynes, in: P. Becher (ed.), *Encyclopedia of Emulsion Technology*, vol. 2, pp. 367–384, Marcel Dekker, New York, 1985.

[3]   S. E. Friberg and I. Kayali, in: M. El-Nokaly and D. Cornell (eds), *Microemulsions and Emulsions in Food*, p. 7, ACS Symposium Series 448, 1991.

[4]   V. Luzzati, in: D. Chapman (ed.), *Biological Membranes*, Academic Press, New York, p. 71 (1968).

[5]   N. Krog and A. P. Borup, J. Sci. Food Agric., **24**, 691 (1973).

[6]   G. Lindblom, K. Larsson, L. Johansson, K. Fontell and S. Forsen, J. Amer. Chem. Soc. **101**, 5465 (1979).

[7]   K. Larsson, K. Fontell and N. Krog, *Chem. Phys. Lipids* **27** (1980), 321.

[8]   E. Pilman, E. Tonberg and K. Lartsson, *J. Dispersion Sci. Technol.* **3** (1982), 335.

[9]   H. Mierovitch and H. A. Scheraga, *Macromolecules* **13** (1980), 1406.

[10] C. Tanford, *Adv. Protein Chem.* **24** (1970), 1.

[11] D. Mobius and R. Miller, Editors, *Proteins at Liquid Interfaces*, Elsevier, Amsterdam, 1998.

[12] P. G. de Gennes, *Scaling Concepts in Polymer Physics*, Corenell University Press, Ithaca, New York, 1979.

[13] K. Larsson, J. Dispersion Sci. Technol., **1**, 267 (1980).

[14] E. Dickinson and P. Walstra (eds.), *Food Colloids and Polymers: Stability and Mechanical Properties*, Royal Society of Chemistry Publication, Cambridge, 1993.

[15] E. D. Goddard and K. P. Ananthapadmanqabhan (eds.). *Polymer-Surfactant Interaction*, CRC Press, Boca Raton, 1992.

[16] P. O. Jansson and S. E. Friberg, *Mol. Cryst. Liq. Cryst.* **34** (1976), 75.

[17] M. M. Breuer, in: P. Becher (ed.), *Encyclopedia of Emulsion Technology*, vol. 2. Ch. 7, Marcel Dekker, New York, 1985.

[18] S. Harry, in: J. B. Wilkinson and R. J. Moore (eds.), *Cosmeticology*, Chemical Publishing, New York, 1981.

[19] S. E. Friberg, *J. Soc. Cosmet. Chem.* **41** (1990), 155.

[20] A. M. Kligman, in: W. Montagna (ed.), *Biology of the Stratum Corneum in Epidermis*, pp. 421–446, Academic Press, New York, 1964.

[21] P. M. Elias, B. E. Brown, P. T. Fritsch, R. J. Gorke, G. M. Goay and R. J. White, *J. Invest. Dermatol.* **73** (1979), 339.

[22] S. E. Friberg and D. W. Osborne, *J. Disp. Sci. Technol.* **6** (1985), 485.

[23] S. C. Vick, Soaps, *Cosmet. Chem. Spec.* **36** (1984).

[24] M. S. Starch, *Drug Cosmet. Ind.* **134** (1984), 38.

[25] Tharwat Tadros, *Applied Surfactants*, Wiley-VCH, Germany, 2005.

[26] R. W. Wahrlow, *Rheological Techniques*, Ellis Horwood Ltd., John Wiley and Sons, New York, 1980.

[27] Tharwat Tadros, *Rheology of Dispersions*, Wiley-VCH, Germany, 2010.

[28] N. Casson, in: C. C. Hill (ed.), *Rheology of Disperse Systems*, p. 84, Pergamon Press, Oxford, 1959.

[29] Tharwat Tadros, *Colloids in Agrochemicals*, Wiley-VCH, Germany, 2009.

[30] T. Foster and W. Von Rybinski, Applications of Emulsions, in: B. P. Binks (ed.), *Modern Aspects of Emulsion Science*, Ch. 12, The Royal Society of Chemistry, Cambridge, 1998.

Index

www.ingramcontent.com/pod-product-compliance
Lightning Source LLC
Chambersburg PA
CBHW061407210326
41598CB00035B/6134